GRAPHING CALCULATOR MANUAL

DARRYL NESTER

Bluffton University

to accompany

TRIGONOMETRY

EIGHTH EDITION

Margaret L. Lial
American River College

John Hornsby
University of New Orleans

David I. Schneider
University of Maryland

PEARSON

Addison Wesley

Boston San Francisco New York
London Toronto Sydney Tokyo Singapore Madrid
Mexico City Munich Paris Cape Town Hong Kong Montreal

Reproduced by Pearson Addison-Wesley from electronic files supplied by the author.

Copyright © 2005 Pearson Education, Inc.
Publishing as Pearson Addison-Wesley, 75 Arlington Street, Boston, MA 02116

ISBN   0-321-22738-7

6 BB 08 07

TABLE OF CONTENTS

This manual contains four chapters, each devoted to one calculator. The table below lists, for each calculator, the topics covered in the introductory section (which contains a brief introduction to using the calculator), followed by the examples from *Trigonometry* which are discussed in this manual. The page numbers in parentheses in the first column—e.g., "Appendix A Example 1 (page 434)"—refer to *Trigonometry*, while the page numbers in the other four columns (under "83/83+/84+," etc.) refer to this manual. For example, information about the TI-86 begins on page 53.

— *continued* —

This graphing calculator manual was written to help owners of Texas Instruments graphing calculators (TI-83 through TI-89) use them to solve problems from *Trigonometry* by Lial, Hornsby, and Schneider. Owners of other TI calculators, like the TI-80, -81, -82, and -92, as well as owners of other brands of graphers, may find the information useful as well, but certainly some of the specific details will not translate directly. In particular, TI-82 users should find many comments in the TI-83/83+/84+ chapter that apply to their calculator, while TI-92 users should read the chapter on the TI-89.

Please contact me with any questions or corrections concerning this material. My web site also contains additional resources for using graphing calculators; I welcome suggestions as to what else would be useful.

This manual was created with TI calculators (of course), the TI-Graph Link and TI-Connect software, Adobe Photoshop, and TEX (Textures from Blue Sky Research).

Darryl Nester
Bluffton University
Bluffton, Ohio
nesterd@bluffton.edu
http://www.bluffton.edu/~nesterd
June 2004

Introduction

The information in this section is essentially a summary of material that can be found in the TI-83 manual. Consult that manual for more details. **All references in this chapter to the TI-83 also apply to the TI-83+ and the TI-84+.**

While the TI-82 and TI-83 differ in some details, in most cases the instructions given in this chapter can be applied (perhaps with slight alteration) to a TI-82. The icon 82 is used to identify significant differences between the two, but some differences (e.g., a slight difference in keystrokes between the two calculators) are not noted. TI-82 users should watch for these comments. Also, see page 26 for information on computing with complex numbers on the TI-82.

1   Power

To power up the calculator, simply press the ON key. This should bring up the "home screen"—a flashing block cursor, and possibly the results of any previous computations that might have been done.

If the home screen does not appear, one may need to adjust the contrast (see the next section).

To turn the calculator off, press 2nd ON (note that the "second function" of ON—written in yellow type above the key—is "OFF"). The calculator will automatically shut off if no keys are pressed for several minutes.

2   Adjusting screen contrast

If the screen is too dark (all black), decrease the contrast by pressing 2nd then pressing and holding ▼. If the screen is too light, increase the contrast by pressing 2nd and then press and hold ▲.

As one adjusts the contrast, the numbers 1 through 9 will appear in the upper right corner of the screen. If the contrast setting reaches 8 or 9, or if the screen never becomes dark enough to see, the batteries should be replaced.

3   Replacing batteries

To replace the four AAA batteries, first turn the calculator off (2nd ON), then remove the back cover, remove and replace each battery, replace the back cover, then turn the calculator on again. (After replacing batteries, one may need to adjust the contrast down as described above.)

4   Basic operations

Simple computations are entered in essentially the same way they would be written. For example, to compute $2 + 17 \times 5$, press $\boxed{2}\boxed{+}\boxed{1}\boxed{7}\boxed{\times}\boxed{5}\boxed{\text{ENTER}}$ (the $\boxed{\text{ENTER}}$ key tells the calculator to act on what has been typed). Standard order of operations (including parentheses) is followed.

```
2+17*5
            87
■
```

The result of the most recently entered expression is stored in Ans, which is typed by pressing $\boxed{\text{2nd}}\boxed{(-)}$ (the word "ANS" appears in yellow above this key). For example, $\boxed{5}\boxed{+}\boxed{\text{2nd}}\boxed{(-)}\boxed{\text{ENTER}}$ will add 5 to the result of the previous computation.

```
2+17*5
            87
5+Ans
            92
■
```

After pressing $\boxed{\text{ENTER}}$, the TI-83 automatically produces Ans if the first key pressed is one which requires a number before it; the most common of these are $\boxed{+}$, $\boxed{-}$, $\boxed{\times}$, $\boxed{\div}$, $\boxed{\wedge}$, $\boxed{x^{-1}}$, $\boxed{x^2}$, and $\boxed{\text{STO}\blacktriangleright}$. For example, $\boxed{+}\boxed{5}\boxed{\text{ENTER}}$ would accomplish the same thing as the keystrokes above (that is, it adds 5 to the previous result).

```
2+17*5
            87
5+Ans
            92
Ans+5
            97
■
```

Pressing $\boxed{\text{ENTER}}$ by itself evaluates the previously typed expression again. This can be especially useful in conjunction with Ans. The screen on the right shows the result of pressing $\boxed{\text{ENTER}}$ a second time.

```
2+17*5
            87
5+Ans
            92
Ans+5
            97
           102
■
```

Several expressions can be evaluated together by separating them with colons ($\boxed{\text{ALPHA}}\boxed{.}$). When $\boxed{\text{ENTER}}$ is pressed, the result of the *last* computation is displayed. The screen shown illustrates the computation $2(5 + 1)^2$.

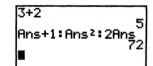

5   Cursors

When typing, the appearance of the cursor indicates the behavior of the next keypress. When the standard cursor (a flashing solid block, ■) is visible, the next keypress will produce its standard action—that is, the command or character printed on the key itself.

If $\boxed{\text{2nd}}\boxed{\text{DEL}}$ is pressed, the TI-83 is placed in INSERT mode and the standard cursor will appear as a flashing underscore. If the arrow keys ($\boxed{\blacktriangle}$, $\boxed{\blacktriangledown}$, $\boxed{\blacktriangleright}$, $\boxed{\blacktriangleleft}$) are used to move the cursor around within the expression, and the TI-83 is placed in INSERT mode, subsequent characters and commands will be inserted in the line at the cursor's position. When the cursor appears as a block, the TI-83 is in DELETE (or OVERWRITE) mode, and subsequent keypresses will replace the character or command at the cursor's position. (When the cursor is at the end of the expression, this is irrelevant.)

The TI-83 will return to DELETE mode when any arrow key is pressed. It can also be returned to DELETE mode by pressing $\boxed{\text{2nd}}\boxed{\text{DEL}}$ a second time.

Pressing $\boxed{\text{2nd}}$ causes an arrow to appear in the cursor: $\boxed{\uparrow}$ (or an underscored arrow). The next keypress will produce its "second function"—the command or character printed in yellow above the key. (The cursor will then return to "standard.") If $\boxed{\text{2nd}}$ is pressed by mistake, pressing it a second time will return the cursor to standard.

Pressing [ALPHA] places the letter "A" in the cursor: Ⓐ (or an underscored "A"). The next keypress will produce the letter or other character printed in green above that key (if any), and the cursor will then return to standard. Pressing [ALPHA] a second time cancels ALPHA mode. Pressing [2nd][ALPHA] "locks" the TI-83 in ALPHA mode, so that all of the following keypresses will produce characters until [ALPHA] is pressed again, or until some menu or second function is accessed.

6   Accessing previous entries ("deep recall")

By repeatedly pressing [2nd][ENTER] ("ENTRY"), previously typed expressions can be retrieved for editing and re-evaluation. Pressing [2nd][ENTER] once recalls the most recent entry; pressing [2nd][ENTER] again brings up the second most recent, etc. The number of previous entries thus displayed varies with the length of each expression (the TI-83 allocates 128 bytes to store previous expressions).

7   Menus

Keys such as [WINDOW], [MATH] and [VARS] bring up a menu screen with a variety of options. The top line of the menu screen gives a collection of submenus (if any), which can be selected with the [◄] and [►] keys. The lower lines list the available commands; these can be selected using the [▲] and [▼] keys and [ENTER], or by pressing the number (or letter) preceding the desired option. Shown is the menu produced by pressing [MATH]; the arrow next to the 7 in the bottom row indicates that there are more options available below.

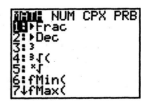

The various commands in these menus are too numerous to be listed here. They will be mentioned as needed in the examples.

8   Variables

The letters A through Z can be used as variables (or "memory") to store numerical values. To store a value, type the number (or an expression) followed by [STO►], then a letter (preceded by [ALPHA] if necessary), then [ENTER]. That letter can then be used in the same way as a number, as demonstrated at right.

Note: The TI-83 interprets 2A as "2 times A"—the "*" symbol is not required (this is consistent with how we interpret mathematical notation). As for order of operations, this kind of multiplication is treated the same as "*" multiplication.

[82] *This latter comment is **not** true of the TI-82; on the TI-82, implied multiplication (such as 2A) is done before other multiplication and division, and even before some other operations, like the square root function ∫. Therefore, for example, the expression 1/2A is evaluated as 1/4 on the TI-82 (assuming that A is 2).*

9   Setting the modes

By pressing the MODE key, one can change many aspects of how the calculator behaves. For most of the examples in this manual, the "default" settings should be used; that is, the MODE screen should be as shown on the right. Each of the options is described below; consult the TI-83 manual for more details. Changes in the settings are made using the arrows keys and ENTER.

The Normal Sci Eng setting specifies how numbers should be displayed. The screen on the right shows the number 12345 displayed in Normal mode (which displays numbers in the range $\pm 9,999,999,999$ with no exponents), Sci mode (which displays all numbers in scientific notation), and Eng mode (which uses only exponents that are multiples of 3). Note: "E" is short for "times 10 to the power," so $1.2345\text{E}4 = 1.2345 \times 10^4 = 1.2345 \times 10000 = 12345$.

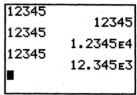

The Float 0123456789 setting specifies how many places after the decimal should be displayed. The default, Float, means that the TI-83 should display all non-zero digits (up to a maximum of 10).

Radian Degree indicates whether angle measurements should be assumed to be in radians or degrees. (A right angle measures $\frac{\pi}{2}$ radians, which is equivalent to $90°$.) Most of the examples in the text include reminders to set the calculator in the appropriate mode, in cases where this is important.

Func Par Pol Seq specifies whether formulas entered into the Y= screen are functions (specifically, y as a function of x), parametric equations (x and y as functions of t), polar equations (r as a function of θ), or sequences (u, v and w as functions of n). The text accompanying this manual uses the first three of these settings.

When plotting a graph, the Connected Dot setting tells the TI-83 whether or not to connect the individually plotted points. Sequential Simul specifies whether individual expressions should be graphed one at a time (sequentially), or all at once (simultaneously).

Real a+bi re^θi specifies how to deal with complex numbers. Real means that only real results will be allowed (unless i is entered as part of a computation)—so that, for example, taking the square root of a negative number produces an error ("NONREAL ANS"). Selecting one of the other two options means that square roots of negative numbers are allowed, and will be displayed in "rectangular" ($a + bi$) or polar ($re^{i\theta}$) format. These two formats are essentially the same as the two used by the textbook. **Note:** The text prefers the term "trigonometric format" rather than "polar format." More information about complex number formats can be found beginning on page 21 of this manual.

[82] *The TI-82 does not support complex numbers, so it does not include this setting. However, some complex computations can be done with a TI-82; see the appendix at the end of this chapter, page 26.*

Finally, the Full Horiz G-T setting allows the option of, for example, showing both the graph and the home screen, as in the screen on the right (this shows a "horizontal split").

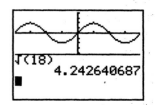

The third option, G-T, has no effect on the home screen display, but will show graphs and tables side by side when GRAPH is pressed.

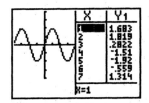

82 *The TI-82 supports only the horizontal split.*

A related group of settings are found in the FORMAT menu (2nd ZOOM). The default settings are shown in the screen on the right, and are generally the best choices for most examples in this book (although the last two settings could go either way). 82 *The TI-82 does not include the* ExprOn ExprOff *option.*

RectGC PolarGC specifies whether graph coordinates should be displayed in rectangular (x, y) or polar (r, θ) format.   Note that this choice is independent of the Func Par Pol Seq mode setting.   The CoordOn CoordOff setting determines whether or not graph coordinates should be displayed. GridOff GridOn specifies whether or not to display a grid of dots on the graph screen, while AxesOn AxesOff and LabelOff LabelOn do the same thing for the axes and labels (y and x) on the axes. ExprOn ExprOff specifies whether or not to display the formula (expression) of the curves on the GRAPH screen when tracing. (This can be useful when more than one graph is displayed.)

10   Setting the graph window

The exact contents of the WINDOW menu vary depending on whether the calculator is in function, parametric, polar, or sequence mode; below are four examples showing the WINDOW menu in each of these modes.

Function mode	Parametric mode	Polar mode	Sequence mode
WINDOW Xmin=-4.7 Xmax=4.7 Xscl=1 Ymin=-3.1 Ymax=3.1 Yscl=1 Xres=1	WINDOW Tmin=0 Tmax=6.2831853… Tstep=.1308996… Xmin=-4.7 Xmax=4.7 Xscl=1 ↓Ymin=-3.1	WINDOW θmin=0 θmax=6.2831853… θstep=.1308996… Xmin=-4.7 Xmax=4.7 Xscl=1 ↓Ymin=-3.1	WINDOW nMin=4 nMax=10 PlotStart=1 PlotStep=1 Xmin=-4.7 Xmax=4.7 ↓Xscl=1

All these menus include the values Xmin, Xmax, Xscl, Ymin, Ymax, and Yscl. When the GRAPH key is pressed, the TI-83 will show a portion of the Cartesian (x-y) plane determined by these values. In function mode, this menu also includes Xres, the behavior of which is described in section 12 of this manual (page 7).   The other settings in the WINDOW screen allow specification of the smallest, largest, and step values of t (for parametric mode) or θ (for polar mode), or initial conditions for sequence mode.

With settings as in the example screens shown above, the TI-83 would display the screen at right: x values from -4.7 to 4.7 (that is, from Xmin to Xmax), and y values between -3.1 to 3.1 (Ymin to Ymax). Since Xscl $=$ Yscl $= 1$, the TI-83 places tick marks on both axes every 1 unit; thus the x-axis ticks are at -4, $-3, \ldots$, 3, and 4, and the y-axis ticks fall on the integers from -3 to 3. This window is called the "decimal" window, and is most quickly set by pressing ZOOM 4.

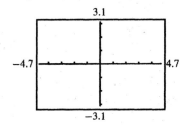

Below are four more sets of WINDOW settings, and the graph screens they produce. Note that the first graph on the left has tick marks every 10 units on both axes. The second window is called the "standard" viewing window, and is most quickly set by pressing $\boxed{\text{ZOOM}}\boxed{6}$. The setting $\text{Yscl} = 0$ in the final graph means that no tick marks are placed on the y-axis.

```
WINDOW
 Xmin=-47
 Xmax=47
 Xscl=10
 Ymin=-31
 Ymax=31
 Yscl=10
 Xres=1
```

```
WINDOW
 Xmin=-10
 Xmax=10
 Xscl=1
 Ymin=-10
 Ymax=10
 Yscl=1
 Xres=1
```

```
WINDOW
 Xmin=-10
 Xmax=100
 Xscl=10
 Ymin=-200
 Ymax=200
 Yscl=50
 Xres=1
```

```
WINDOW
 Xmin=-.5
 Xmax=1
 Xscl=.1
 Ymin=-10
 Ymax=10
 Yscl=0
 Xres=1
```

11   The graph screen

The TI-83 screen is made up of an array of square dots (pixels) with 63 rows and 95 columns. All the pixels in the leftmost column have x-coordinate Xmin, while those in the rightmost column have x-coordinate Xmax. The x-coordinate changes steadily across the screen from left to right, which means that the coordinate for the nth column (counting the leftmost column as column 0) must be $\text{Xmin} + n\Delta\text{X}$, where $\Delta\text{X} = (\text{Xmax} - \text{Xmin})/94$. Similarly, the nth row of the screen (counting up from the bottom row, which is row 0) has y-coordinate $\text{Ymin} + n\Delta\text{Y}$, where $\Delta\text{Y} = (\text{Ymax} - \text{Ymin})/62$.

Note: In (horizontal) split screen mode, $\Delta\text{Y} = (\text{Ymax} - \text{Ymin})/30$. In G-T (vertical split screen) mode, $\Delta\text{X} = (\text{Xmax} - \text{Xmin})/46$ and $\Delta\text{Y} = (\text{Ymax} - \text{Ymin})/50$.

It is not necessary to memorize the formulas for ΔX and ΔY. Should they be needed, they can be determined by pressing $\boxed{\text{GRAPH}}$ and then the arrow keys. When pressing $\boxed{\blacktriangleright}$ or $\boxed{\blacktriangleleft}$ successively, the displayed x-coordinate changes by ΔX; meanwhile, when pressing $\boxed{\blacktriangle}$ or $\boxed{\blacktriangledown}$, the y-coordinate changes by ΔY. Alternatively, the values 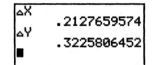 can be found by pressing $\boxed{\text{VARS}}\boxed{1}\boxed{8}\boxed{\text{ENTER}}$ (for ΔX) or $\boxed{\text{VARS}}\boxed{1}\boxed{9}\boxed{\text{ENTER}}$ (for ΔY). This produces results like those shown on the right.

In the decimal window $\text{Xmin} = -4.7$, $\text{Xmax} = 4.7$, $\text{Ymin} = -3.1$, $\text{Ymax} = 3.1$, note that $\Delta\text{X} = 0.1$ and $\Delta\text{Y} = 0.1$. Thus, the individual pixels on the screen represent x-coordinates -4.7, -4.6, -4.5, ..., 4.5, 4.6, 4.7 and y-coordinates -3.1, -3, -2.9, ..., 2.9, 3, 3.1. This is where the decimal window gets its name.

Windows for which $\Delta\text{X} = \Delta\text{Y}$, such as the decimal window, are called square windows. Any window can be made square be pressing $\boxed{\text{ZOOM}}\boxed{5}$. To see the effect of a square window, observe the two pairs of graphs below. In each pair, the first graph is on the standard window, and the second is on a square window (after pressing $\boxed{\text{ZOOM}}\boxed{5}$). The first pair shows the lines $y = 2x - 3$ and $y = 3 - \frac{1}{2}x$; note that on the square window, these lines look perpendicular (as they should). The second pair shows a circle centered at the origin with a radius of 8. On the standard window, this looks like an oval since the screen is wider than it is tall. (The reason for the gaps in the circle will be addressed in the next section.)

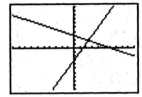 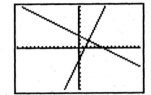 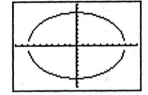 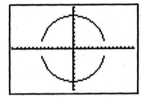

12   Graphing a function

This introductory section only addresses creating graphs in function mode. Procedures for creating parametric and polar graphs are very similar; they are covered beginning on page 23 of this manual, in material related to Chapter 8 of the text. Sequence plotting is not covered in the text.

To see the graph of $y = 2x - 3$, begin by entering the formula into the calculator. This is done on the $\boxed{Y=}$ screen of the calculator. Select one of the variables Y₁, Y₂, ..., and enter the formula. If other y variables have formulas, either erase them (by positioning the cursor on that line and pressing $\boxed{CLEAR}$) or position the cursor

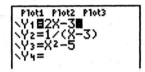

on the equals sign "=" for that line and press $\boxed{ENTER}$ (this has the effect of "unhighlighting" the equals sign, which tells the TI-83 not to graph that formula). Additionally, if any of Plot1, Plot2 or Plot3 are highlighted, move the cursor up until it is on that plot and press $\boxed{ENTER}$. In the screen on the right, only Y₁ will be graphed.

The next step is to choose a viewing window; see the previous section for more details on this. This example uses the standard window ($\boxed{ZOOM}\boxed{6}$).

Finally, press $\boxed{GRAPH}$, and the line should be drawn. In order to produce this graph, the TI-83 considers 95 values of x, ranging from Xmin to Xmax in steps of ΔX (assuming that Xres $= 1$; see below for other possibilities). For each value of x, it computes the corresponding value of y, then plots that point (x, y) and (if the calculator is in Connected mode) draws a line between this point and the previous one.

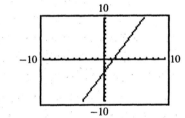

If Xres is set to 2, the TI-83 will only compute y for every other x value; that is, it uses a step size of $2\Delta X$. Similarly, if Xres is 3, the step size will be $3\Delta X$, and so on. Setting Xres higher causes graphs to appear faster (since fewer points are plotted), but for some functions, the graph may look "choppy" if Xres is too large, since detail is sacrificed for speed.

Note: If the line does not appear, or the TI-83 reports an error, double-check all the previous steps. Also, check the mode settings (discussed in section 9, page 4).

Once the graph is visible, the window can be changed using $\boxed{WINDOW}$ or $\boxed{ZOOM}$. Pressing the $\boxed{TRACE}$ key brings up the "trace cursor," and displays the x- and y-coordinates for various points on the line as the $\boxed{\blacktriangleleft}$ and $\boxed{\blacktriangleright}$ keys are pressed. Tracing beyond the left or right columns causes the TI-83 to adjust the values of Xmin and Xmax and redraw the graph.

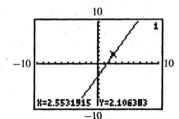

To graph the function

$$y = \frac{1}{x - 3},$$

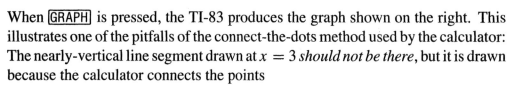

enter that formula into the $\boxed{\text{Y=}}$ screen (note the use of parentheses). As before, this example uses the standard viewing window.

When $\boxed{\text{GRAPH}}$ is pressed, the TI-83 produces the graph shown on the right. This illustrates one of the pitfalls of the connect-the-dots method used by the calculator: The nearly-vertical line segment drawn at $x = 3$ *should not be there*, but it is drawn because the calculator connects the points

$\qquad x = 2.9787234, \; y = -46.99999$ and $\; x = 3.1914894, \; y = 5.2222223.$

Calculator users must learn to recognize these flaws in calculator-produced graphs.

The graph of a circle centered at the origin with radius 8 (shown on the square window $\boxed{\text{ZOOM}}\boxed{6}\boxed{\text{ZOOM}}\boxed{5}$) shows another problem that arises from connecting the dots. When $x = -8.064516$, y is undefined, so no point is plotted (that is, there is no point on this circle that has x-coordinate less than -8, or greater than 8). The next point plotted on the upper half of the circle is $x = -7.741935$ and $y = 2.0155483$; since no point had been plotted for the previous x-coordinate, this is not connected to anything, so there appears to be a gap between the circle and the x-axis. The calculator is not "smart" enough to know that the graph should extend from -8 to 8.

One additional feature of graphing with the TI-83 is that each function can have a "style" assigned to its graph. The symbol to the left of Y₁, Y₂, etc. indicates this style, which can be changed by pressing $\boxed{\triangleleft}$ until the cursor is over the symbol, then pressing $\boxed{\text{ENTER}}$ to cycle through the options. These options are shown on the right (with brief descriptive names); complete details are in the TI-83 manual.

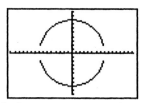

---

13   Adding programs to the TI-83

The TI-83's capabilities can be extended by downloading or entering programs into the calculator's memory. Instructions for writing a program are beyond the scope of this manual, but programs written by others and downloaded from the Internet (or obtained as printouts) can be transferred to the calculator in one of three ways:

1. If one TI-83 already has a program, it can be transferred to another using the calculator-to-calculator link cable. To do this, first make sure the cable is firmly inserted in both calculators. On the sending calculator, press $\boxed{\text{2nd}}\boxed{\text{X,T,Θ,n}}$ (LINK), then $\boxed{3}$, and then select (by using the $\boxed{\blacktriangle}$ and $\boxed{\blacktriangledown}$ keys and $\boxed{\text{ENTER}}$) the program(s) to be transferred. Now press the $\boxed{\blacktriangleright}$ to bring up the TRANSMIT submenu. *Before* pressing $\boxed{\text{ENTER}}$ on the sending calculator, prepare the receiving calculator by pressing $\boxed{\text{2nd}}$ $\boxed{\text{X,T,Θ,n}}\boxed{\blacktriangleright}\boxed{\text{ENTER}}$, and *then* press $\boxed{\text{ENTER}}$ on the sending calculator.

2. If a computer with the TI-Graph Link is available, and the program file is on that computer (e.g., after having been downloaded from the Internet), the program can be transferred to the calculator using the TI Graph Link software. This transfer is done in a manner similar to the calculator-to-calculator transfer described above; specific instructions can be found in the documentation that accompanies the Graph Link software. (They are not given here because of slight differences between platforms and software versions.)

3. View a listing of the program and type it in manually. (**Note:** Even if the TI-Graph Link cable is not available, the Graph Link software can be used to view program listings on a computer.) While this is the most tedious method, studying programs written by others can be a good way to learn programming. To enter a program, start by choosing [PRGM][◀][1] ("Create New"), then type a name for the new program (like "QUADFORM" or "MIDPOINT")—note that the TI-83 is automatically put into ALPHA mode. Then type each command in the program, and press [2nd][MODE] (QUIT) to return to the home screen when finished.

To run the program, make sure there is nothing on the current line of the home screen, then press [PRGM], select the number or letter of the program (a sample screen is shown), and press [ENTER]. If the program was entered manually (option 3 above), errors may be reported; in that case, choose GOTO, correct the mistake and try again.

Programs can be found at many places on the Internet, including:

- `http://www.awl.com/lhs`—the Web site for the text;

- `http://www.bluffton.edu/~nesterd`—the Web site of the author of this manual;

- `http://tifaq.calc.org`—A "Frequently Asked Questions" page maintained by Ray Kremer; and

- `http://www.ticalc.org`.

Examples

Here are the details for using the TI-83 for several of the examples from the textbook. Also given are the keystrokes necessary to produce some of the commands shown in the text's examples. In some cases, some suggestions are made for using the calculator more efficiently.

We first consider examples from the text's Appendices, as the calculator techniques they illustrate are useful throughout the text.

Throughout this section, it is assumed that the textbook is available for reference. The problems from the text are not restated here, and there are frequent references to the calculator screens shown in the text.

Appendix A   Example 1   (page 434)   Solving a Linear Equation

Here is a general discussion of how to use the TI-83 to solve (or confirm solutions for) nearly any equation. We will show multiple approaches for solving the equation $\frac{1}{2}x - 6 = \frac{3}{4}x - 9$. (The answer is $x = 12$.) These procedures can be adapted for any equation, including the one from this example, or those found throughout the text.

There are two graphical methods that can be used to confirm this solution. The first is the **intersection** method. To begin, set up the TI-83 to graph the left side of the equation as Y1, and the right side as Y2. **Note:** Putting the fractions in parentheses ensures no mistakes with order of operations. This is not crucial for  the TI-83, but is a good practice because some other models give priority to implied multiplication. See section 8 of the introduction, page 3.

We are looking for an x value that will make the left and right sides of this equation equal to each other, which corresponds to the x-coordinate of the point of intersection of these two graphs.

Next, select a viewing window which shows the point of intersection; we use $[-15, 15] \times [-10, 10]$ for this example. The TI-83 can automatically locate this point using the CALC menu ([2nd][TRACE]). Choose option 5 (intersect), use [▲], [▼] and [ENTER] to specify which two functions to use (in this case, the only two being displayed), and then use [◄] or [►] to specify a guess. After pressing [ENTER], the TI-83 will try to find an intersection of the two graphs. The screens below illustrate these steps.

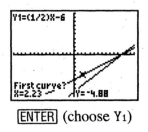

[2nd][TRACE][5]

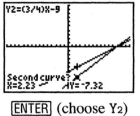

[ENTER] (choose Y1)

[ENTER] (choose Y2)

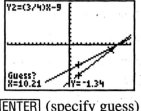

[ENTER] (specify guess)

The final result of this process is the screen shown on the right. The x-coordinate of this point of intersection is calculated to 14 digits of accuracy, so if the solution were some less "convenient" number (say, $\sqrt{3}$ or $1/\pi$), we would have an answer that would be accurate enough for nearly any computation.

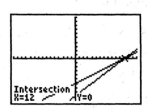

Note: An approximation for the point of intersection can be found simply by moving the TRACE cursor as near the intersection as possible. The amount of error can be minimized by "zooming in" on the graph. This is the only method available for graphing calculators such as the TI-81.

The second graphical approach is to use the x-**intercept method**, which seeks the x-coordinate of the point where a graph crosses the x-axis. Specifically, we want to know where the graph of Y₁−Y₂ crosses the x-axis, where Y₁ and Y₂ are as defined above. This is because the equation $\frac{1}{2}x - 6 = \frac{3}{4}x - 9$ can only be true when $\frac{1}{2}x - 6 - \left(\frac{3}{4}x - 9\right) = 0$.

To find this x-intercept, begin by defining Y₃=Y₁−Y₂ on the Y= screen. We could do this by re-typing the formulas entered for Y₁ and Y₂, but having typed those formulas once, it is more efficient to do this as shown on the right. To type Y₁ and Y₂, press VARS ▶ 1 to access the Y-VARS:FUNCTION menu. Note that Y₁ and Y₂ have been "de-selected" so that they will not be graphed (see section 12 of the introduction, page 7).

We must first select a viewing window which shows the x-intercept; we again use $[-15, 15] \times [-10, 10]$. The TI-83 can automatically locate this point by choosing option 2 from the CALC (2nd TRACE) menu. The TI-83 prompts for left and right bounds and a guess, then attempts to locate the zero between the given bounds. (Provided there is only one zero between the bounds, and the function is "well-behaved"—meaning it has some nice properties like continuity—the calculator will find it.) The screens below illustrate these steps.

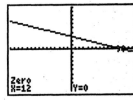

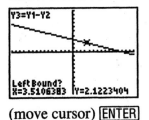

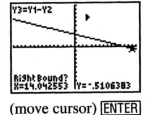

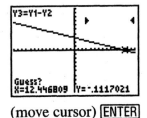

(move cursor) ENTER   (move cursor) ENTER   (move cursor) ENTER   Here is the result.

The TI-83 also offers some non-graphical approaches to solving this equation (or confirming a solution): As illustrated on the right, the TI-83's solve function attempts to find a value of X that makes the given expression equal to 0, given a guess (10, in this case). The entry shown use of the fact that Y₁ and Y₂ have been

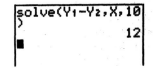

defined as the left and right sides of this equation; if that had not been the case, the same results could have been attained by entering (e.g.) solve((1/2)X-6-((3/4)X-9),X,10).

The solve command is accessed through the function catalog: Press 2nd 0, then 4 (which brings up the commands beginning with "t"). Press ▲ until the triangle cursor (▸) points to solve(, then press ENTER. For some equations, the "guess" value does not affect the result, but for some equations—especially those with more than one solution—different guesses might produce different results. Full details on how to use this function can be found in the TI-83 manual.

Since it is difficult to access this command, it would be a good idea to either use deep recall (see page 3) if solve is needed several times in a row, or to use the "interactive solver" built-in to the TI-83. This feature is found by pressing MATH 0.

If Solver has not been used before, the first screen below (a) appears, asking for an expression involving at least one variable. The TI-83 will attempt to find values of that variable which make the expression equal to 0. The expression for this example is shown in screen (b). Pressing ENTER brings up screen (c). Type the guess (this example uses 5), then press ALPHA ENTER (SOLVE) and the TI-83 seeks a solution near 5, which is reported in screen (d).

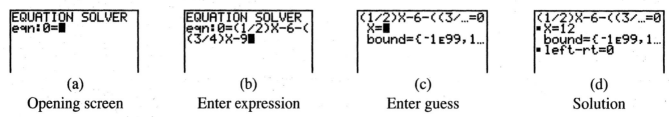

(a)	(b)	(c)	(d)
Opening screen	Enter expression	Enter guess	Solution

If Solver *has* been used before, MATH 0 jumps directly to screen (c). To change the expression, press ▲.

Solver can also be used with equations containing more than one variable; simply provide values for all but one variable, then place the cursor on the line containing the variable for which a value is needed and press ALPHA ENTER.

82 *The TI-82 does not include this interactive solver; the* solve *function is accessed with* MATH 0.

Note: In this example, we learned how the TI-83 can be used to support an analytic solution. But the TI-83 and any other graphing calculator also can be used for solving problems when an analytic solution is **not** possible—that is, when one cannot solve an equation "algebraically." This is often the case in many "real-life" applications, and is one of the best arguments for the use of graphing calculators.

Appendix A   Example 7   (page 438)   Solving a Linear Inequality

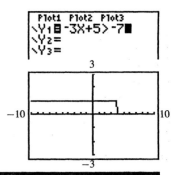

Shown is one way to visualize the solution to inequalities like these. The inequality symbols >, <, ≥, ≤ are found in the TEST menu (2nd MATH). The TI-83 responds with 1 when a statement is true and 0 for false statements. This is why the graph of Y_1 appears as it does: For values of x less than 4, the inequality is true (and so Y_1 equals 1), and for $x \geq 4$, the inequality is false (and Y_1 equals 0). Note that this picture does *not* help one determine what happens when $x = 4$.

Appendix A   Example 8   (page 439)   Solving a Three-Part Inequality

To see a "picture" of this inequality like the one shown for the previous example requires a little additional work. Setting Y_1=-2<5+3X≤20 will *not* work correctly. Instead, one must enter either

Y_1=(-2<5+3X)(5+3X≤20)   or Y_1=(-2<5+3X) and (5+3X≤20)

which will produce the desired results: A function that equals 1 between $-\frac{7}{3}$ and 5, and equals 0 elsewhere. ("and" is found in the TEST:LOGIC menu, 2nd MATH ▶.)

Appendix B   Example 2   (page 444)   Using the Midpoint Formula

The TI-83 can do midpoint computations nicely by putting coordinates in a list—
that is, using braces ([2nd][(] and [2nd][)]) instead of parentheses. When adding two
lists, the calculator simply adds corresponding elements, so the two x-coordinates
are added, as are the y-coordinates. Dividing by 2 completes the task.

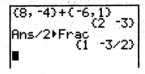

Appendix C   Example 5   (page 455)   Using Function Notation

The TI-83 will do *some* computations in function notation. Simply enter the
function in Y₁ (or any other function). Return to the home screen, and press [VARS]
[▶] (Y-VARS). Option 1 allows access to all the function (y) variables; parametric
and polar variables are also available (as options 2 and 3). Select the appropriate
function (Y₁ in this case), then type (2) to evaluate that function with the input 2.
The result shown on the screen agrees with (a) in the text.

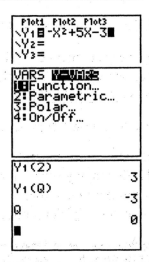

The TI-83 is less helpful if asked to evaluate expressions like those given in
Example 5(b). Note, for example, that Y₁(Q) produces the result -3, rather than
the desired result $-Q^2 + 5Q - 3$. When asked to evaluate an expression using
a variable (like Q), the TI-83 simply substitutes the current value of that variable,
which in this case was 0.

The table features of the TI-83 allow another method of computing function values. To use these features,
begin by entering the formula on the [Y=] screen, as one would to create a graph. (The highlighted equals
signs determine which formulas will be displayed in the table, just as they do for graphs.)

Next, press [2nd][WINDOW] to access the TABLE SETUP screen. The table will display y
values for given values of x. The TblStart value sets the lowest value of x, while
ΔTbl determines the "step size" for successive values of x. These two values are
only used if the Indpnt option is set to Auto—this means, "automatically generate
the values of the independent variable (x)." For this usage, Indpnt should be set to Ask. (The Depend
option should almost always be set to Auto; if it is set to Ask, the y values are not displayed until [ENTER]
is pressed.)

When the TABLE SETUP options are set satisfactorily, press [2nd][GRAPH] to produce
the table. In the screen shown, values of Y₁ are shown for the three input values
0, 1, and 2. (These three input values had to be individually entered.)

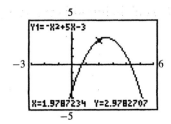

To get graphical confirmation, enter the appropriate formula for Y₁ and graph
in any window that includes $x = 2$. Press [TRACE], then press the [▶] and [◀] keys
to change the value of x. It may not be possible to make x exactly equal to 2 in
this manner (see section 12 of the introduction, page 7), but rough confirmation
that $f(2) = 3$ can be found by observing that y is close to 3 when x is close
to 2.

The TI-83 makes it possible to trace to any real number value for x between Xmin and Xmax. Simply type a number or expression (like $1/\pi$ or $\sqrt{}(2)$) while in TRACE mode. The number appears at the bottom of the window in a larger font size than the TRACE coordinates. Pressing ENTER causes the TRACE cursor to jump to that x-coordinate. This same result can be achieved using option 1 (value) from the 2nd TRACE (CALC) menu. This latter approach is also available on the TI-82.

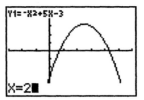 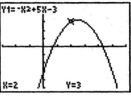

Section 1.1   Example 2   (page 4)   Calculating with Degrees, Minutes, and Seconds

---

Section 1.1   Example 3   (page 4)   Converting between Decimal Degrees
                                                            and Degrees, Minutes, and Seconds

---

See section 9 of this chapter's introduction (page 4) for information about selecting Degree mode. The alternative to putting the calculator in Degree mode is to use the degree symbol (2nd MATRX 1) following each angle measure.

The degrees and minutes symbols, and the ▸DMS operator (which causes an angle to be displayed in degrees, minutes, and seconds, rather than as a decimal), are all found in the ANGLE (2nd MATRX) menu, shown on the right. The "seconds" symbol is simply the double quotes, entered as ALPHA +.

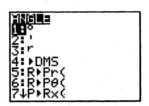

Here we see the addition for Example 2(a). Note that, similar to what is seen in the screen accompanying Example 3 in the text, we must use ▸DMS to see the result of the addition in degrees, minutes, and seconds.

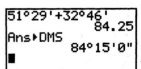

[82] *The TI-82 uses only the minutes symbol "'" for entering angles in degrees, minutes, and seconds (output uses all the symbols). For example, to convert* $74°8'14''$ *to decimal degrees, one would type* 74'8'12'.

Section 1.2   Example 1   (page 10)   Finding Angle Measures

See page 10 for general information about solving equations using the TI-83. (Of course, one must use knowledge from geometry to obtain the equation in the first place.)

Section 1.3   Example 4   (page 23)   Finding Function Values of Quadrantal Angles

---

The alternative to putting the calculator in Degree mode is to use the degree symbol (2nd MATRX 1) following each angle measure; e.g., enter sin(90°) rather than just sin(90). (However, see the comment in the next example.)

Since the cotangent, secant, and cosecant functions are the reciprocals of the tangent, cosine, and sine, they can be entered as (e.g.) 1/sin(90). (This is illustrated in Figure 31 on page 27 of the text.) Note, though, that this will not properly compute $\cot 90°$, since 1/tan(90) produces a domain error. Entering $\cot x$ as $\cos x / \sin x$ will produce the correct result at 90°.

One might guess that the other three trigonometric functions are accessed with [2nd] followed by [SIN], [COS], or [TAN] (which produce, e.g., sin^{-1}). This is **not** what these functions do; in this case, the exponent -1 does not mean "reciprocal," but instead indicates that these are inverse functions (which are discussed in Sections 2.3 and 6.1 of the text). The text comments on this distinction at the bottom of page 27.

Section 2.3   Example 1   (page 62)   Finding Function Values with a Calculator

The screen on the right shows how computations like the secant in part (b) and the cotangent in (c) can be entered on a single line. These computations were done in Degree mode.

```
sin(49°12'
           .7569950557
(cos(97.977))⁻¹
           -7.205879213
■
```

The screen on the right (with computations done in Radian mode) illustrates a somewhat unexpected behavior: Even if an angle is entered in DMS format, the TI-83 assumes that the angle is in radians. In order to remedy this, either put the calculator in Degree mode, or use the degree symbol ([2nd][MATRX][1]) as was done in the second entry.

```
sin(49°12'
           -.875022579
sin(49°12'°
           .7569950557
■
```

[82] *On a TI-82, the computation in part (a) would be entered* `sin 49'12'`.

Section 2.3   Example 2   (page 62)   Using an Inverse Trigonometric Function to Find an Angle

The sin^{-1} ("inverse sine," or "arcsine") function is [2nd][SIN], while cos^{-1} is [2nd][COS]. The computations shown in Figure 16 in the text were done in Degree mode; the screen on the right shows the result when done in Radian mode. Note that in the second entry, an attempt was made to get the TI-83 to report the result in degrees (by placing the degree symbol at the end of the entry), but this does not have the desired result.

```
sin⁻¹(.9677091705
)
           1.315978256
sin⁻¹(.9677091705
)°
           .0229681535
■
```

Section 2.4   Example 1   (page 69)   Solving a Right Triangle Given an Angle and a Side

---

Section 2.4   Example 2   (page 70)   Solving a Right Triangle Given Two Sides

For problems like these, the Ans variable can be used to avoid loss of accuracy from rounding off intermediate results. Shown are computations for Example 1. (Be sure the TI-83 is in Degree mode. Also, recall that the arcsine (or inverse sine) function is [2nd][SIN] *—not* [SIN][x⁻¹].)

```
12.7sin(34°30')
           7.193359209
√(12.7²-Ans²)
           10.4664026
sin⁻¹(Ans/12.7)▶D
MS
               55°30'0"
■
```

Section 2.5   Example 4   (page 79)   Solving a Problem Involving Angles of Elevation

The TI-83 can automatically locate the intersection of two graphs using the CALC menu ([2nd][TRACE]). This feature was previously illustrated on page 10, but we repeat the description here: Choose option 5 (intersect), use [▲], [▼] and [ENTER] to specify which two functions to use (in this case, the only two being displayed), and then use [◄] or [►] to specify a guess. After pressing [ENTER], the TI-83 will try to find an intersection of the two graphs. The screens below illustrate these steps; the final result is the screen shown as Figure 32 of the text. The guessing step in the fourth screen below is not crucial in this case, since the calculator would locate the intersection even if a very poor guess was given.

   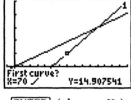   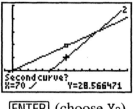   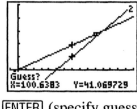

[2nd][TRACE][5]   [ENTER] (choose Y₁)   [ENTER] (choose Y₂)   [ENTER] (specify guess)

Section 3.1   Example 1   (page 95)   Converting Degrees to Radians

The number π is available as [2nd][^], and the degree symbol is [2nd][MATRX][1]. With the calculator in Radian mode (see page 4), entering $45°$ causes the TI-83 to automatically convert to radians.

A useful technique to aid in recognizing when an angle is a multiple of π is to divide the result by π. This approach is illustrated in the screen on the right, showing that $45°$ is $\pi/4$ radians, and $30°$ is $\pi/6$ radians. This screen also makes use of the ▸Frac command ([MATH][1]), which simply means "display the result of this computation as a fraction, if possible."

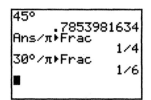

An alternative to using the degree symbol is to store $\pi/180$ in the calculator variable D (see page 3). Then typing, for example, 45D [ENTER] will multiply 45 by $\pi/180$. This approach will work regardless of whether the calculator is in Degree or Radian mode. (A value stored in a variable will remain there until it is replaced by a new value.)

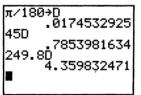

Section 3.1   Example 2   (page 95)   Converting Radians to Degrees

With the TI-83 in Degree mode (see page 4), the radian symbol (a superscripted r), produced with [2nd][MATRX][3], will automatically change a radian angle measurement to degrees.

Alternatively, with the value $180/\pi$ stored in the calculator variable R (see page 3), typing $(9\pi/4)$R [ENTER] will convert from radians to degrees regardless of whether the calculator is in Degree or Radian mode. (The same result can be achieved by *dividing by* the calculator variable D as defined in the previous example.)

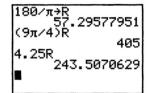

Section 3.3   Example 4   (page 112)   Finding a Number Given Its Circular Function Value

The arccosine (or inverse cosine) function is [2nd][COS] —*not* [COS][x⁻¹]. Likewise, $\tan^{-1}$ is [2nd][TAN].

Section 4.1   Example 1   (page 135)   Graphing $y = a\sin x$

Note that the TI-83 must be in Radian mode in order to produce the desired graph. See page 8 for information on setting the thickness of a graph.

The graphs in the text, and those below, are shown in the "trig viewing window," described on the top of page 135. This window can be selected automatically by pressing ZOOM 7. (If the TI-83 is in Degree mode, Xmin and Xmax will be ±360 instead of ±2π.)

It is possible to distinguish between the two graphs without having them drawn using different styles by using the TRACE feature. On the right, the trace cursor is on graph 2—that is, the graph of $y_2 = \sin x$.

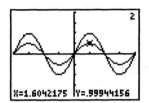

Alternatively, when ExprOn is selected on the TI-83's graph format screen, the expression is displayed in the upper left corner of the screen.

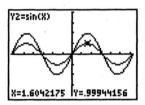

82 *The TI-82 has neither graph styles nor* ExprOn, *so the* TRACE *feature mentioned above is the only option.*

Section 4.2   Example 6   (page 151)   Modeling Temperature with a Sine Function

---

Given a set of data pairs (x, y), the TI-83 can produce a scatter diagram (like the points shown in Figure 19) and can find various formulas (including linear and quadratic, as well as more complex formulas like a sine function) that approximate the relationship between x and y. These formulas are called "regression formulas."

82 *The TI-82 will find many types of regression formulas, but does not support sine regression. It may be possible to find a program to perform a sine regression (see section 13 of the introduction, page 8).*

The first step is to enter the data into the TI-83. This is done by pressing STAT, then choosing option 1 (Edit).

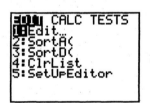

This brings up the list-editing screen. (If the column headings are not L₁, L₂, L₃ as shown on the right, press STAT 5 ENTER to reset the statistics [list] editor to its default. Then return to the list-editing screen.) Enter the month number into the first column (L₁) and the temperature into the second column (L₂). If either column already contains data, the DEL key can be used to delete numbers one at a time, or—to delete the whole column at once—press the ▲ key until the cursor is at the top of the column (on L₁ or L₂) and press CLEAR ENTER. Make sure that both columns contain the same number of entries. (For the scatter diagram shown in the text, enter the temperatures twice—24 pairs of number altogether.)

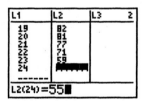

To produce the scatter diagram, press 2nd Y= 1 and make the settings for a STAT PLOT shown on the right.

Now check that nothing else will be plotted: Press [Y=] and make sure that the only thing highlighted is Plot1. If Plot2 or Plot3 is highlighted, use the arrow keys to move the cursor up to that plot, then press [ENTER].

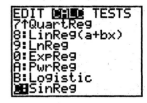

Finally, set up the viewing window as shown in Figure 19 of the text—or press [ZOOM][9] (ZoomStat), which automatically adjusts the window to show all the data in the plot. This should produce a plot like that shown in the text.

Note: When finished with a STAT PLOT, it is a good idea to turn all statistics plots off so that the TI-83 will not attempt to display them the next time [GRAPH] is pushed. This is most easily done by executing the PlotsOff command, using the key sequence [2nd][Y=][4][ENTER].

To find the regression equation, press [STAT][▶][ALPHA][PRGM] to choose option C: SinReg from the CALC statistics submenu. This will place the command SinReg on the home screen.

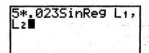

To tell the TI-83 where to find *x* and *y*, press [2nd][1][,][2nd][2], which adds "L₁,L₂" to the command. Pressing [ENTER] should then produce the screen similar to Figure 21(a). (That screen was produced with the calculator set to display two decimal places.) **Note:** Adding "L₁,L₂" is not absolutely necessary, since the TI-83 will assume that *x* is in L₁ and *y* is in L₂ unless told otherwise. In other words, "SinReg" (by itself) works the same as "SinReg L₁,L₂".

If this process produces an error message, it will likely either be a DIM MISMATCH (meaning that the two lists L₁ and L₂ have different numbers of entries) or a SYNTAX error, probably because the SinReg command was not on a line by itself. The screen on the right, for example, produces a syntax error.

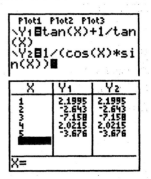

Section 4.3   Example 1   (page 158)   Graphing $y = a \sec bx$

Section 4.3   Example 2   (page 159)   Graphing $y = a \csc(x - d)$

See section 9 of the introduction (page 4) for information about Connected versus Dot mode. The function in Example 1 can be entered as Y₁=2cos(X/2)⁻¹ (or as shown in the text). The function in Example 2 can be entered as Y₁=(3/2)(1/sin(X−π/2)) or Y₁=3/(2sin(X−π/2)). A reminder: sin⁻¹ ([2nd][SIN]) is *not* the cosecant function.

Section 5.1   Example 3   (page 185)   Rewriting an Expression in Terms of Sine and Cosine

The top screen on the right shows how these expressions are entered on the [Y=] screen. As an alternative to graphing these two functions, the TI-83's table feature (see page 13 of this manual) can be used: If the *y* values are the same for a reasonably large sample of *x* values, one can be fairly sure (though not certain) that the two expressions are equal. To make this approach more reliable, be sure to choose *x* values that are not, for example, all multiples of π.

| Section 5.3 | Example 1 | (page 198) | Finding Exact Cosine Function Values |

| Section 5.4 | Example 1 | (page 206) | Finding Exact Sine and Tangent Function Values |

The TI-83 can graphically and numerically support exact value computations such as $\cos 15° = \frac{\sqrt{6}+\sqrt{2}}{4}$. Starting with a graph of Y₁=cos(X), the TI-83 makes it possible to trace to any real number value for x between Xmin and Xmax. Simply type a number or expression (like 1/π or √

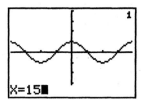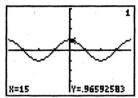

(2)) while in TRACE mode. The number appears at the bottom of the window in a larger font size than the TRACE coordinates. Pressing [ENTER] causes the TRACE cursor to jump to that x-coordinate. This same result can be achieved using option 1 (value) from the [2nd][TRACE] (CALC) menu.

[82] *This latter approach is also available on the TI-82.*

Alternatively, a table of values like those shown here can be used to find the value of $\cos 15°$. Of course, the screen shown in the text in support of Example 1(b) shows the other part of this process: Computing the decimal value of $(\sqrt{6}+\sqrt{2})/4$ and observing that it agrees with those found here.

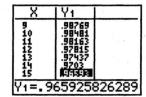

Note that Example 1 in Section 5.6 (page 222) shows that $\cos 15°$ can also be written as $\frac{\sqrt{2+\sqrt{3}}}{2}$.

| Section 6.1 | Example 1 | (page 238) | Finding Inverse Sine Values |

| Section 6.1 | Example 2 | (page 239) | Finding Inverse Cosine Values |

Of course, it is not necessary to graph $y = \sin^{-1} x$ or $y = \cos^{-1} x$ to find these values; one can simply enter, e.g., sin⁻¹(1/2) on the home screen. The first entry of the screen on the right shows what happens when the calculator is in Degree mode; note that the result is not in $[-\pi/2, \pi/2]$. With the calculator in Radian mode, results similar to those in the text are found, and the method employed on page 16 of this manual (in the discussion of text Example 1 from Section 3.1) confirms that these values are $\pi/6$ and $3\pi/4$.

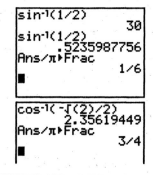

| Section 6.1 | Example 4 | (page 243) | Finding Inverse Function Values with a Calculator |

Note that the answer given for (b), 109.499054°, overrepresents the accuracy of that value. A typical rule for doing computations involving decimal values (like −0.3541) is to report only as many digits in the result as were present in the original number—in this case, four. This means the reported answer should be "about 109.5°," and in fact, any angle θ between about 109.496° and 109.501° has a cotangent which rounds to −0.3541. (See the discussion of significant digits on page 68 of the text.)

Section 6.2   Example 6   (page 253)   Describing a Musical Tone from a Graph

Section 6.3   Example 5   (page 259)   Analyzing Pressures of Upper Harmonics

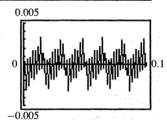

Note that the calculator screens shown in Figures 23–26 illustrate the importance of choosing a "good" viewing window. If we choose the wrong vertical scale (Ymin and Ymax), we might not be able to see the graph at all—it might be squashed against the x-axis. If we make the window too wide—that is, if Xmax minus Xmin is too large—we might see the "wrong" picture, like the one on the right (for Example 5 from Section 6.3): We seem to see six periods in this view, when in fact there are 44. Each of the six "pseudo-cycles" is made of parts of seven or eight full periods.

This observation—that a periodic function, viewed at fixed intervals, can appear to be a *different* periodic function—is the same effect that causes wagon wheels to appear to run backwards in old movies.

Section 7.4   Example 1   (page 307)   Finding Magnitude and Direction Angle

The R▸Pr(and R▸Pθ(conversion functions (illustrated in Figure 26) are in the TI-83's ANGLE menu ([2nd] [MATRX]) as options 5 and 6. R▸Pθ(x, y) returns an angle in degrees if the TI-83 is in Degree mode.

"R" and "P" stand for "rectangular" and "polar" coordinates. Rectangular coordinates are the familiar x and y values. Polar coordinates, described in Section 8.5 of the text, are r (which corresponds to the magnitude of a vector) and θ (the direction angle).

Section 7.4   Example 2   (page 307)   Finding Horizontal and Vertical Components

The P▸Rx(and P▸Ry(conversion functions are in the ANGLE menu ([2nd] [MATRX]) as options 7 and 8. Note that the screen shown in the text was created with the calculator set to display only one digit after the decimal, and in Degree mode. If the TI-83 is in Radian mode, the degree symbol (also in the ANGLE menu) can specify that the angle is in degrees.

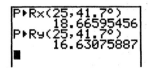

"R" and "P" stand for "rectangular" and "polar" coordinates. Rectangular coordinates are the familiar x and y values. Polar coordinates, described in Section 8.5 of the text, are r (which corresponds to the magnitude of a vector) and θ (the direction angle).

Finally, observe that this conversion could be done by typing 25cos(41.7) and 25sin(41.7); which approach to use is a matter of personal preference.

Section 7.4   Example 5   (page 309)   Performing Vector Operations

---

Section 7.4   Example 6   (page 310)   Finding Dot Products

Vector operations might be performed with a program (see page 8), or the vectors can be entered as lists, as Figure 33 shows. If a vector is to be used more than once, it may be useful to store it in one of the TI-83's list variables L_1, L_2, ..., L_6. For dot products, one can use the sum command, found in the LIST:MATH menu ([2nd][STAT][◄]).

```
{-2,1}→L₁:{4,3}→
L₂:4L₁-3L₂
              {-20  -5}
{2,3}{4,-1}
              {8  -3}
sum(Ans)
                    5
■
```

Section 8.1   Example 1   (page 333)   Writing $\sqrt{-a}$ as $i\sqrt{a}$

Section 8.1   Example 4   (page 334)   Finding Products and Quotients Involving Negative Radicands

The TI-83 will not do these computations unless it is first put in a+bi mode; see page 4. (One could also use re^θi mode, but this would report complex numbers in "polar format," rather than the format used in the text.)

[82] *The TI-82 will not do computations with complex results. See the appendix to the chapter (page 26) for information about how to "fake" these computations.*

Section 8.1   Example 5   (page 335)   Adding and Subtracting Complex Numbers

---

Section 8.1   Example 6   (page 336)   Multiplying Complex Numbers

---

Section 8.1   Example 7   (page 337)   Simplifying Powers of i

---

Section 8.1   Example 8   (page 338)   Dividing Complex Numbers

The character " i " is [2nd][.]. Although not completely necessary, it is a good idea to put the TI-83 in a+bi mode; see page 4. (Even in Real mode, the TI-83 will display results for computations in which i is entered directly; it only complains if asked to find even roots of negative numbers.)

[82] *See the appendix to this chapter (page 26) for information about doing these computations on a TI-82.*

Section 8.2   Example 2   (page 342)   Converting from Trigonometric Form to Rectangular Form

While the TI-83 must be in a+bi or re^θi mode in order to compute (e.g.) square roots of negative numbers, it will perform computations involving i ([2nd][.]) even while in Real mode (see page 4).

Section 8.2   Example 3   (page 343)   Converting from Rectangular Form to Trigonometric Form

The TI-83 can convert from rectangular to trigonometric
form in several ways. Aside from the ANGLE ([2nd][MATRX])
menu commands R▸Pr(and R▸Pθ(, shown in the text,
one can use the MATH:CPX menu commands angle(and
abs(. The angle(command gives the angle θ in radians

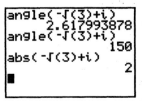
or degrees (depending on the mode setting), and the abs(function (also found in the MATH:NUM menu)
gives the modulus r. These functions are shown on page 347 of the text.

Furthermore, with the calculator placed in re^θi mode (see page 4), the conversion can be done all at once, as the screen on the right shows (the TI-83 was
also in Degree mode, so the angles are reported in degrees). As the name of the
mode suggests, the values of r and θ can be read directly from the output: r is the
leading coefficient, and θ is in the exponent.

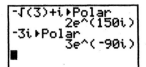

Additionally, using the ▸Polar command (option 7 in the MATH CPX menu; [MATH]
[▸][▸][7]), and with the TI-83 in **any** mode (Real, a+bi, or re^θi), the output is
displayed in this polar format, as the screen on the right shows. Note: Polar format
is equivalent to trigonometric format because of the (fairly deep) mathematical
fact that $e^{i\theta} = \cos\theta + i\sin\theta = \operatorname{cis}\theta$.

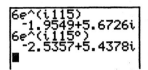

Section 8.2   Example 4   (page 344)   Converting Between Trigonometric and Rectangular Forms
Using Calculator Approximations

Aside from entering the expression in (a) as it is shown in the text (using the cos
and sin functions), one can use the polar format (re^θi), but note that *regardless
of the TI-83's* Degree/Radian *mode setting,* all such computations are done with the
angle θ in radians. Thus, the first result on the right is incorrect, because the
cosine and sine were computed for 115 radians (this is true even if the TI-83 is in Degree mode). The second
entry, with the degree symbol included, gives the correct result, but *this will only work if the TI-83 is in*
Radian *mode!*

Section 8.3   Example 1   (page 348)   Using the Product Theorem

Section 8.3   Example 2   (page 349)   Using the Quotient Theorem

Section 8.3   Example 3   (page 349)   Using the Product and Quotient Theorems with a Calculator

The TI-83's polar complex format makes these computations very easy to enter.
(But, be sure to put parentheses around the denominator for Example 2; otherwise,
the TI-83 will multiply by cis 150° instead of dividing.)

Section 8.4   Example 1   (page 353)   Finding a Power of a Complex Number

The screen on the right shows several options for computing $(1 + i\sqrt{3})^8$ with the TI-83 (in a+bi mode). The first is fairly straightforward, but the reported result shows the complex part of the answer $(128i\sqrt{3})$ given in decimal form. For an "exact" answer, the TI-83's polar format (re^θi) can be used. The TI-83 was in Degree mode for the first of the two polar-format answers, and in Radian mode for the second.

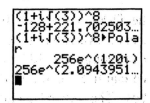

Section 8.5   Example 3   (page 361)   Examining Polar and Rectangular Equations
of Lines and Circles

Section 8.5   Example 4   (page 363)   Graphing a Polar Equation (Cardioid)

Section 8.5   Example 5   (page 363)   Graphing a Polar Equation (Rose)

Section 8.5   Example 6   (page 364)   Graphing a Polar Equation (Lemniscate)

To produce these polar graphs, the TI-83 should be set to Degree and Polar modes (see the screen on the right). In this mode, the [Y=] key allows entry of up to six polar equations (r as a function of θ). One could also use Radian mode, adjusting the values of θmin, θmax, and θstep accordingly (e.g., use 0, 2π, and $\pi/30$ instead of 0, 360, and 5).

For the cardioid, rose, and lemniscate, the window settings shown in the text show these graphs on "square" windows (see section 11 of the introduction, page 6), so one can see how their proportions compare to those of a circle.

For the cardioid, the value of θstep does not need to be 5, although that choice works well for this graph. Too large a choice of θstep produces a graph with lots of sharp "corners," like the one shown on the right (drawn with θstep=30). Setting θstep too small, on the other hand, produces a smooth graph, but it is drawn very slowly. Sometimes it may be necessary to try different values of θstep to choose a good one.

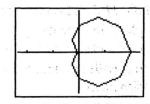

The lemniscate can be drawn by setting θmin=0 and θmax=180, or θmin=-45 and θmax=45. In fact, with θ ranging from -45 to 225, the graph of r$_1$=$\sqrt{\,}$(cos(2θ)) (alone) will produce the entire lemniscate. (θstep should be about 5.)

The rose can be produced by setting θmin=0 and θmax=360, or using any 360°-range of θ values (with θstep about 5).

Section 8.5   Example 7   (page 365)   Graphing a Polar Equation (Spiral of Archimedes)

To produce this graph on the viewing window shown in the text, the TI-83 must be in Radian mode. (In Degree mode, it produces the same shape, but magnified by a factor of $180/\pi$ —meaning that the viewing window needs to be larger by that same factor.)

Section 8.6   Example 1   (page 371)   Graphing a Plane Curve Defined Parametrically

Place the TI-83 in Parametric mode, as the screen on the right shows. In this mode, the Y= key allows entry of up to six pairs of parametric equations (x and y as functions of t). No graph is produced unless both functions in the pair are entered and selected (that is, both equals signs are highlighted).

The value of Tstep does not need to be 0.05, although that choice works well for this graph. Too large a choice of Tstep produces a less-smooth graph, like the one shown on the right (drawn with Tstep=1). Setting Tstep too small, on the other hand, produces a smooth graph, but it is drawn very slowly. Sometimes it may be necessary to try different values of Tstep to choose a good one.

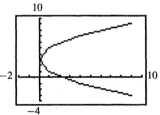

Section 8.6   Example 3   (page 372)   Graphing a Plane Curve Defined Parametrically

This curve can be graphed in Degree mode with Tmin=0 and Tmax=360, or in Radian mode with Tmax=2π. In order to see the proportions of this ellipse, it might be good to graph it on a square window. This can be done most easily by pressing ZOOM 5. On a TI-83, initially with the window settings shown in the text, this would result in the window $[-6, 6] \times [-4, 4]$.

Section 8.6   Example 5   (page 373)   Graphing a Cycloid

The TI-83 *must* be in Radian mode in order to produce this graph.

Section 8.6   Example 6   (page 374)   Simulating Motion with Parametric Equations

Section 8.6   Example 8   (page 375)   Analyzing the Path of a Projectile

Parametric mode is particularly nice for analyzing motion, because one can picture the motion by watching the calculator create the graph, or by pressing TRACE and watching the motion of the trace cursor. (When tracing in parametric mode, the ▶ and ◀ keys increase and decrease the value of t, and the trace cursor shows the location (x, y) at time t.) Figure 40 illustrates tracing on the projectile path in Example 8. Note that the value of t changes by $\pm$Tstep each time ▶ or ◀ is pressed, so obviously the choice of Tstep affects which points can be traced.

The TI-83's graph styles (see page 8) can be useful, too. The screen on the right shows the three paths for Example 6 being plotted in "ball path" style.

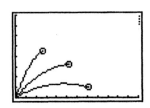

Section 9.1   Example 4   (page 393)   Using a Property of Exponents to Solve an Equation

Below the calculator screen shown in the text, the caption refers to "the *x*-intercept method of solution." This and other methods for solving equations were described beginning on page 10 of this manual.

Section 9.1   Example 11   (page 397)   Using Data to Model Exponential Growth

The scatter diagram in Figure 10(a) and the "exponential regression" (ExpReg, option 0 in the STAT:CALC menu) in Figure 12 can be reproduced by adapting the procedures described on page 17 of this manual.

The two calculator screens in Figure 11 use the "intersection method" of solving equations; see page 10 for a description.

Section 9.3   Example 1   (page 414)   Finding pH

For (a), the text shows $-\log(2.5*10^{\wedge}(-4))$, but this could also be entered as shown on the first line of the screen on the right, since "E" (produced with 2nd ,) and "$*10^{\wedge}$" are nearly equivalent. The two are not completely interchangeable, however; in particular, in part (b), "$10^{\wedge}$" **cannot** be replaced with "E", because "E" 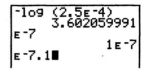 is only valid when followed by an *integer*. That is, E-7 produces the same result as $10^{\wedge}-7$, but the last line shown on the screen produces a syntax error.

(Incidentally, "$10^{\wedge}$" is 2nd LOG, but 1 0 ^ produces the same results.)

Appendix: Simulating complex numbers with a TI-82

The TI-82 does not support complex numbers; however, it can be made (using matrices) to add, subtract, multiply, and divide complex numbers in rectangular format. Here are the details:

Enter two matrices in the TI-82 by reproducing the screen on the right. The " [" character is 2nd×, and "]" is 2nd-. Enter " [A] " as MATRX1, and " [B] " as MATRX2 (do not use 2nd× and 2nd- for these brackets). The screen is shown just before pressing ENTER.

```
[[1,0][0,1]]→[A]

              [[1  0]
               [0  1]]
[[0,1][-1,0]]→[B
]■
```

The matrix [A] stands for "1", and [B] stands for "*i*." To enter $2 - 3i$, for example, type 2[A]-3[B] (**not** just 2-3[B]). Addition and subtraction are simply performed; the screen on the right shows the computation of the addition problem $(3+5i)+(6-2i)$. To obtain the answer $-9+3i$ — simply read the first row of the resulting matrix; the first number is the real part, and the second is the imaginary part.

```
3[A]+5[B]
              [[3   5]
               [-5  3]]
Ans+(6[A]-2[B])
              [[9   3]
               [-3  9]]
■
```

Multiplication is no more complicated than addition and subtraction. Shown below are the calculator entries and outputs for some sample multiplication problems.

```
(2[A]-3[B])(3[A]
+4[B])
              [[18  -1]
               [1   18]]
■
```

```
(5[A]-4[B])(7[A]
-2[B])
              [[27  -38]
               [38  27]]
■
```

```
(6[A]+5[B])(6[A]
-5[B])
              [[61  0 ]
               [0   61]]
■
```

```
(4[A]+3[B])²
              [[7    24]
               [-24  7 ]]
■
```

$$(2-3i)(3+4i)$$
$$=18-i$$

$$(5-4i)(7-2i)$$
$$=27-38i$$

$$(6+5i)(6-5i)$$
$$=61$$

$$(4+3i)^2$$
$$=7+24i$$

For division problems, do not use the ÷ key. Instead, *multiply* by the inverse (x^{-1}) of the denominator.

```
(3[A]+2[B])(5[A]
-[B])⁻¹▶Frac
     [[1/2   1/2]
      [-1/2  1/2]]
■
```

```
(4[A]+2[B])(3[A]
-[B])⁻¹
              [[1   1]
               [-1  1]]
■
```

```
3[A]*[B]⁻¹
              [[0   -3]
               [3   0 ]]
■
```

```
(2[A]+[B])((A]-
2[B])^3)⁻¹▶Frac
     [[-4/25   -3/25]
      [3/25    -4/25]]
■
```

$$\frac{3+2i}{5-i}=\frac{1}{2}+\frac{1}{2}i$$

$$\frac{4+2i}{3-i}=1+i$$

$$\frac{3}{i}=-3i$$

$$\frac{2+i}{(1-2i)^3}=-\frac{4}{25}-\frac{3}{25}i$$

Introduction

The information in this section is essentially a summary of material that can be found in the TI-85 manual. Consult that manual for more details.

1   Power

To power up the calculator, simply press the ON key. This should bring up the "home screen"—a flashing block cursor, and possibly the results of any previous computations that might have been done.

If the home screen does not appear, one may need to adjust the contrast (see the next section).

To turn the calculator off, press 2nd ON (note that the "second function" of ON—written in yellow type above the key—is "OFF"). The calculator will automatically shut off if no keys are pressed for several minutes.

2   Adjusting screen contrast

If the screen is too dark (all black), decrease the contrast by pressing 2nd then pressing and holding ▾. If the screen is too light, increase the contrast by pressing 2nd and then press and hold ▴.

As one adjusts the contrast, the numbers 1 through 9 will appear in the upper right corner of the screen. If the contrast setting reaches 8 or 9, or if the screen never becomes dark enough to see, the batteries should be replaced.

3   Replacing batteries

To replace the four AAA batteries, first turn the calculator off (2nd ON), then remove the back cover, remove and replace each battery, replace the back cover, then turn the calculator on again. (After replacing batteries, one may need to adjust the contrast down as described above.)

4   Basic operations

Simple computations are entered in essentially the same way they would be written. For example, to compute $2 + 17 \times 5$, press [2][+][1][7][×][5][ENTER] (the [ENTER] key tells the calculator to act on what has been typed). Standard order of operations (including parentheses) is followed.

```
2+17*5
                   87
■
```

The result of the most recently entered expression is stored in Ans, which is typed by pressing [2nd][(-)] (the word "ANS" appears in yellow above this key). For example, [5][+][2nd][(-)][ENTER] will add 5 to the result of the previous computation.

```
2+17*5
                   87
5+Ans
                   92
■
```

After pressing [ENTER], the TI-85 automatically produces Ans if the first key pressed is one which requires a number before it; the most common of these are [+], [−], [×], [÷], [∧], [x^2], and [STO►]. For example, [+][5][ENTER] would accomplish the same thing as the keystrokes above (that is, it adds 5 to the previous result).

```
2+17*5
                   87
5+Ans
                   92
Ans+5
                   97
■
```

Pressing [ENTER] by itself evaluates the previously typed expression again. This can be especially useful in conjunction with Ans. The screen on the right shows the result of pressing [ENTER] a second time.

```
2+17*5
                   87
5+Ans
                   92
Ans+5
                   97
                  102
■
```

Several expressions can be evaluated together by separating them with colons ([2nd][.]). When [ENTER] is pressed, the result of the *last* computation is displayed. The screen shown illustrates the computation $2(5 + 1)^2$.

```
3+2
                    5
Ans+1:Ans²:2 Ans
                   72
■
```

5   Cursors

When typing, the appearance of the cursor indicates the behavior of the next keypress. When the standard cursor (a flashing solid block, ■) is visible, the next keypress will produce its standard action—that is, the command or character printed on the key itself.

If [2nd][DEL] is pressed, the TI-85 is placed in INSERT mode and the standard cursor will appear as a flashing underscore. If the arrow keys ([▲], [▼], [▶], [◀]) are used to move the cursor around within the expression, and the TI-85 is placed in INSERT mode, subsequent characters and commands will be inserted in the line at the cursor's position. When the cursor appears as a block, the TI-85 is in DELETE (or OVERWRITE) mode, and subsequent keypresses will replace the character(s) at the cursor's position. (When the cursor is at the end of the expression, this is irrelevant.)

The TI-85 will return to DELETE mode when any arrow key is pressed. It can also be returned to DELETE mode by pressing [2nd][DEL] a second time.

Pressing [2nd] causes an arrow to appear in the cursor: ◨ (or an underscored arrow). The next keypress will produce its "second function"—the command or character printed in yellow above the key. (The cursor will then return to "standard.") If [2nd] is pressed by mistake, pressing it a second time will return the cursor to standard.

Pressing ALPHA places the letter "A" in the cursor: ▣ (or an underscored "A"). The next keypress will produce the letter or other character printed in blue above that key (if any), and the cursor will then return to standard. Pressing 2nd ALPHA puts the calculator in lowercase ALPHA mode, changing the cursor to ▤ and producing the lowercase version of a letter. Pressing ALPHA twice (or 2nd ALPHA ALPHA) "locks" the TI-85 in ALPHA (or lowercase ALPHA) mode, so that all of the following keypresses will produce characters until ALPHA is pressed again.

6   Accessing previous entries

By pressing 2nd ENTER ("ENTRY"), the last expression can be recalled to the home screen for editing and/or re-execution. This can be useful if a fairly complicated sequence of commands must be executed several times, with slight variations each time.

7   Menus

Keys such as STAT, GRAPH and 2nd 7 (MATRX) bring up a menu line at the bottom of the screen with a variety of options. These options can be selected by pressing one of the function keys (F1, F2, . . . , F5). If the menu ends with a small triangle ("▶"), it means that more options are available in this menu, which can be viewed by pressing MORE. Shown is the menu produced by pressing 2nd × (MATH).

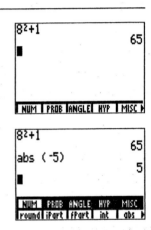

This screen shows the result of pressing F1 (the "NUM" option, which lists a variety of numerical functions). Note that the MATH menu still appears (with NUM highlighted) and the bottom line now lists the functions available in this sub-menu—including, for example, the absolute value function (abs), which is accessed by pressing F5. The command line abs (-5) was typed by pressing F5 ((-) 5) ENTER.

This manual will use (e.g.) MATH:NUM to indicate commands accessed through menus like this. Sometimes the keypresses will be included as well; for this example, it would be 2nd × F1 F5.

The various commands in these menus are too numerous to be listed here. They will be mentioned as needed in the examples.

One last comment is worthwhile, however. Some functions that may be used frequently are buried several levels deep in the menus, and may take many keystrokes to access. Worse, the location of the function might be forgotten (is it MATH:NUM or MATH:MISC?), necessitating a search through the menus. It is useful to remember three things:

- Any command can be typed one letter at a time, in either upper- or lowercase; e.g., ALPHA ALPHA LOG SIN 6 (-) will type the letters "ABS ", which has the same effect as 2nd × F1 F5.

- Any command can be found in the CATALOG menu (2nd CUSTOM). Since the commands appear in alphabetical order, it may take some time to locate the desired function (although abs is easy to find!). Pressing any letter key brings up commands starting with that letter (it is not necessary to press ALPHA first); e.g., pressing , shows commands starting with "P."

- Frequently used commands can be placed in the CUSTOM menu, and will then be available simply by pressing [CUSTOM]. To do this, scroll through the CATALOG to find the desired function, then press [F3] (CUSTM) followed by one of [F1]–[F5] to place that command in the CUSTOM menu. In the screen shown, [F1] was pressed, so that pressing [CUSTOM][F1] will type "Solver(." The commands in the MATH:ANGLE menu ([2nd][×][F3]), used frequently for problems in this text, could be made more accessible by placing them in this menu.

8   Variables

The uppercase letters A through Z, as well as some (but not all) lowercase letters, and also sequences of letters (like "High" or "count") can be used as variables (or "memory") to store numerical values. To store a value, type the number (or an expression) followed by [STO▸], then a letter or letters (note that the TI-85 automatically goes into ALPHA mode when [STO▸] is pressed), then [ENTER]. That variable name can then be used in the same way as a number, as demonstrated at right.

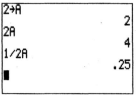

Note: The TI-85 interprets 2A as "2 times A"—the "∗" symbol is not required (this is consistent with how we interpret mathematical notation). However, the various Texas Instruments calculators are not consistent in how they interpret an expression like 1/2A. The TI-82 and TI-85 perform "implied multiplication" *before* the division (see the screen on the right), while the other (newer) models follow the standard order of operations. Owners of the TI-82 and TI-85 need to aware of this quirk.

9   Setting the modes

By pressing [2nd][MORE] (MODE), one can change many aspects of how the calculator behaves. For most of the examples in this manual, the "default" settings should be used; that is, the MODE screen should be as shown on the right. Each of the options is described below; consult the TI-85 manual for more details. Changes in the settings are made using the arrows keys and [ENTER].

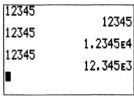

The `Normal Sci Eng` setting specifies how numbers should be displayed. The screen on the right shows the number 12345 displayed in `Normal` mode (which displays numbers in the range $\pm999,999,999,999$ with no exponents), `Sci` mode (which displays all numbers in scientific notation), and `Eng` mode (which uses only exponents that are multiples of 3). Note: "E" is short for "times 10 to the power," so $1.2345\text{E}4 = 1.2345 \times 10^4 = 1.2345 \times 10000 = 12345$.

The `Float 012345678901` setting specifies how many places after the decimal should be displayed (the 0 and 1 at the end mean 10 and 11 decimal places). The default, `Float`, means that the TI-85 should display all non-zero digits (up to a maximum of 12).

`Radian Degree` indicates whether angle measurements should be assumed to be in radians or degrees. (A right angle measures $\frac{\pi}{2}$ radians, which is equivalent to $90°$.) Most of the examples in the text include reminders to set the calculator in the appropriate mode, in cases where this is important.

`RectC PolarC` specifies whether complex numbers should be displayed in rectangular or polar format. These two formats are essentially the same as the two used by the textbook. **Note:** The text prefers the term "trigonometric format" rather than "polar format." More information about complex number formats can be found beginning on page 47 of this manual.

`Func Pol Param DifEq` specifies whether formulas to be graphed are functions (y as a function of x), polar equations (r as a function of θ), parametric equations (x and y as functions of t), or differential equations ($Q'(t)$ as a function of Q and t). The text accompanying this manual uses the first three of these settings.

The `RectV CylV SphereV` setting indicates the default display format for vectors (see page 46 of this manual).

The other two mode settings deal with issues that are beyond the scope of the textbook, and are not discussed here.

A group of settings related to the graph screen are found by pressing GRAPH MORE F3 (GRAPH:FORMT). The default settings are shown in the screen on the right, and are generally the best choices for most examples in this book (although the last setting could go either way).

`RectGC PolarGC` specifies whether graph coordinates should be displayed in rectangular (x, y) or polar (r, θ) format. Note that this choice is independent of the `Func Pol Param DifEq` mode setting.

The `CoordOn CoordOff` setting determines whether or not graph coordinates should be displayed.

When plotting a graph, the `DrawLine DrawDot` setting tells the TI-85 whether or not to connect the individually plotted points. (The text refers to these as Connected and Dot modes.) `SeqG SimulG` specifies whether individual expressions should be graphed one at a time (sequentially), or all at once (simultaneously).

`GridOff GridOn` specifies whether or not to display a grid of dots on the graph screen, while `AxesOn AxesOff` and `LabelOff LabelOn` do the same thing for the axes and labels (y and x) on the axes.

---

10   Setting the graph window

Pressing GRAPH F2 brings up the RANGE settings. (**Note:** These are more appropriately called WINDOW settings; they have nothing to do with the "range" of a function, meaning the set of possible output values.) The exact contents of the RANGE (WINDOW) menu vary depending on whether the calculator is in function, polar, parametric, or DifEq mode; below are four examples showing this menu in each of these modes.

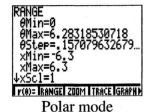

| Function mode | Polar mode | Parametric mode | DifEq mode |

All these menus include the values xMin, xMax, xScl, yMin, yMax, and yScl. When GRAPH F5 (GRAPH) is pressed, the TI-85 will show a portion of the Cartesian (x-y) plane determined by these values. The other settings in this screen allow specification of the smallest, largest, and step values of θ (for polar mode) or t (for parametric mode), or initial conditions for the differential equation.

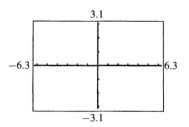

With settings as in the example screens shown above, the TI-85 would display the screen at right: x values from -6.3 to 6.3 (that is, from xMin to xMax), and y values between -3.1 to 3.1 (yMin to yMax). Since xScl $=$ yScl $= 1$, the TI-85 places tick marks on both axes every 1 unit; thus the x-axis ticks are at -6, -5, ..., 5, and 6, and the y-axis ticks fall on the integers from -3 to 3. This window is called the "decimal" window, and is most quickly set by pressing GRAPH F3 (ZOOM) MORE F4 (ZDECM).

Note: If the graph screen has a menu on the bottom (like that shown on the right), possibly obscuring some important part of the graph, it can be removed by pressing CLEAR. The menu can be restored later by pressing EXIT.

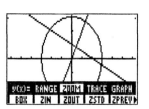

Below are four more sets of window settings, and the graph screens they produce. Note that the first graph on the left has tick marks every 10 units on both axes. The second window is called the "standard" viewing window, and is most quickly set by pressing GRAPH F3 (ZOOM) F4 (ZSTD). The setting yScl $= 0$ in the final graph means that no tick marks are placed on the y-axis.

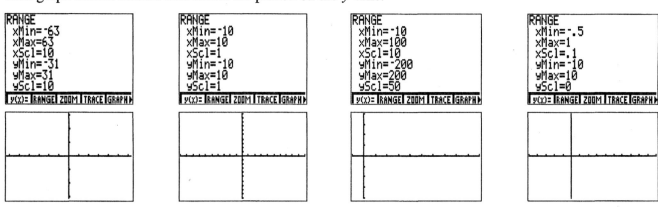

11   The graph screen

The TI-85 screen is made up of an array of rectangular dots (pixels) with 63 rows and 127 columns. All the pixels in the leftmost column have x-coordinate xMin, while those in the rightmost column have x-coordinate xMax. The x-coordinate changes steadily across the screen from left to right, which means that the coordinate for the nth column (counting the leftmost column as column 0) must be xMin $+ n\Delta x$, where $\Delta x = (\text{xMax} - \text{xMin})/126$. Similarly, the nth row of the screen (counting up from the bottom row, which is row 0) has y-coordinate yMin $+ n\Delta y$, where $\Delta y = (\text{yMax} - \text{yMin})/62$.

It is not necessary to memorize the formulas for Δx and Δy. Should they be needed, they can be determined by pressing [GRAPH][F5] and then the arrow keys. When pressing [▶] or [◀] successively, the displayed x-coordinate changes by Δx; meanwhile, when pressing [▲] or [▼], the y-coordinate changes by Δy. Alternatively, the values can be found by typing "Δx" and "Δy" on the home screen; this

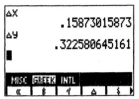

is most easily done by pressing [2nd][0][F2][F4] to access the CHAR:GREEK menu and type the "Δ" character, then typing lowercase x or y. This produces results like those shown on the right; the CHAR:GREEK menu remains on the bottom of the screen.

In the decimal window xMin $= -6.3$, xMax $= 6.3$, yMin $= -3.1$, yMax $= 3.1$, note that $\Delta x = 0.1$ and $\Delta y = 0.1$. Thus, the individual pixels on the screen represent x-coordinates $-6.3, -6.2, -6.1, \ldots, 6.1,$ $6.2, 6.3$ and y-coordinates $-3.1, -3, -2.9, \ldots, 2.9, 3, 3.1$. This is where the decimal window gets its name.

It happens that the pixels on the TI-85 screen are about 1.2 times taller than they are wide, so if $\Delta y/\Delta x$ is approximately 1.2 (the exact value is $1.19565\ldots$), the window will be a "square" window (meaning that the scales on the x- and y-axes are equal). For example, the decimal window (with $\Delta y/\Delta x = 1$) is not square, so that one unit on the x-axis is not the same length as one unit on the y-axis. (Specifically, one y-axis unit is about 20% longer than one x-axis unit.)

Any window can be made square be pressing [GRAPH][F3] (ZOOM) [MORE][F2] (ZSQR). To see the effect of a square window, observe the two pairs of graphs below. In each pair, the first graph is on the standard window, and the second is on a square window (after choosing ZOOM:ZSQR). The first pair shows the lines $y = 2x - 3$ and $y = 3 - \frac{1}{2}x$; note that on the square window, these lines look perpendicular (as they should). The second pair shows a circle centered at the origin with a radius of 8. On the standard window, this looks like an oval since the screen is wider than it is tall. (The reason for the gaps in the circle will be addressed in the next section.)

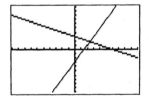 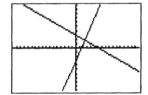 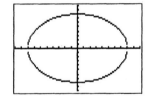 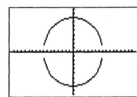

12   Graphing a function

This introductory section only addresses creating graphs in function mode. Procedures for creating parametric and polar graphs are very similar; they are covered beginning on page 49 of this manual, in material related to Chapter 8 of the text.

To see the graph of $y = 2x - 3$, begin by entering the formula into the calculator. This is done by pressing [GRAPH][F1] to access the "y equals" screen of the calculator. Enter the formula as y1 (or any other yn); note that the letter x can be typed by pressing [F1] or [x-VAR] (as well as [2nd][ALPHA][+]). If another y variable has a formula, position the cursor on that line and press either [F4] (DELf—to delete the

function) or [F5] (SELCT). The latter has the effect of toggling the "highlighting" for the equals sign "=" for that line (an "unhighlighted" equals sign tells the TI-85 not to graph that formula). In the screen on the right, only y1 will be graphed.

The next step is to choose a viewing window (or "range," in the terminology of the TI-85). See the previous section for more details on this. This example uses the standard window ([GRAPH][F3][F4]).

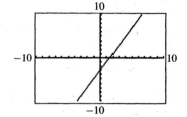

If the graph has not been displayed, press [GRAPH][F5], and the line should be drawn. In order to produce this graph, the TI-85 considers 127 values of x, ranging from xMin to xMax in steps of Δx. For each value of x, it computes the corresponding value of y, then plots that point (x, y) and (if the calculator is in Connected [DrawLine] mode) draws a line between this point and the previous one.

Note: If the line does not appear, or the TI-85 reports an error, double-check all the previous steps. Also, check the mode settings (discussed in section 9, page 30).

Once the graph is visible, the window can be changed using [F2] (RANGE) or [F3] (ZOOM). Pressing [F4] (TRACE) brings up the "trace cursor," and displays the x- and y-coordinates for various points on the line as the [◄] and [►] keys are pressed. Tracing beyond the left or right columns causes the TI-85 to adjust the values of xMin and xMax and redraw the graph.

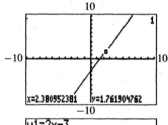

To graph the function

$$y = \frac{1}{x-3},$$

enter that formula into the "y equals" screen (note the use of parentheses). As before, this example uses the standard viewing window.

For this function, the TI-85 produces the graph shown on the right. This illustrates one of the pitfalls of the connect-the-dots method used by the calculator: The nearly-vertical line segment drawn at $x = 3$ *should not be there*, but it is drawn because the calculator connects the points

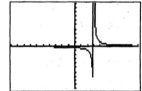

$$x = 2.85714, y = -6.99999 \text{ and } x = 3.01587, y = 62.99999.$$

Calculator users must learn to recognize these flaws in calculator-produced graphs.

The graph of a circle centered at the origin with radius 8 (shown on the square window ZOOM:ZSTD - ZOOM:ZSQR) shows another problem that arises from connecting the dots. When $x = -8.093841$, y is undefined, so no point is plotted (that is, there is no point on this circle that has x-coordinate less than -8, or greater than 8). The next point plotted on the upper half of the circle is $x = -7.824046$ and $y = 1.668619$; since no point had been plotted for the previous x-coordinate, this is not connected to anything, so there appears to be a gap between the circle and the x-axis. The calculator is not "smart" enough to know that the graph should extend from -8 to 8.

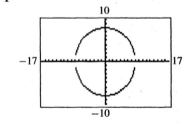

13   Adding programs to the TI-85

The TI-85's capabilities can be extended by downloading or entering programs into the calculator's memory. Instructions for writing a program are beyond the scope of this manual, but programs written by others and downloaded from the Internet (or obtained as printouts) can be transferred to the calculator in one of three ways:

1. If one TI-85 already has a program, it can be transferred to another using the calculator-to-calculator link cable. To do this, first make sure the cable is firmly inserted in both calculators. On the sending calculator, press [2nd][x-VAR] (LINK), then [F1][F2] (SEND:PRGM), and then select (by using the [▲] and [▼] keys and [F2]) the program(s) to be transferred. *Before* pressing [F1] (SEND) on the sending calculator, prepare the receiving calculator by pressing [2nd][x-VAR][F2], and *then* press [F1] on the sending calculator.

2. If a computer with the TI-Graph Link is available, and the program file is on that computer (e.g., after having been downloaded from the Internet), the program can be transferred to the calculator using the TI Graph Link software. This transfer is done in a manner similar to the calculator-to-calculator transfer described above; specific instructions can be found in the documentation that accompanies the Graph Link software. (They are not given here because of slight differences between platforms and software versions.)

3. View a listing of the program and type it in manually. (**Note:** Even if the TI-Graph Link cable is not available, the Graph Link software can be used to view program listings on a computer.) While this is the most tedious method, studying programs written by others can be a good way to learn programming. To enter a program, start by choosing [PRGM][F2] (EDIT), then type a name for the new program (up to eight letters, like "QuadForm" or "Midpoint")—note that the TI-85 is automatically put into ALPHA mode. Then type each command in the program, and press [2nd][EXIT] (QUIT) to return to the home screen when finished.

To run the program, make sure there is nothing on the current line of the home screen, then press [PRGM][F1], select the program using one of the keys [F1]–[F5] and [MORE] (a sample screen is shown; only the first four to six characters of each program name are shown), and press [ENTER]. If the program was entered manually (option 3 above), errors may be reported; in that case, choose GOTO, correct the mistake and try again.

Programs can be found at many places on the Internet, including:

- http://www.awl.com/lhs—the Web site for the text;

- http://www.bluffton.edu/~nesterd—the Web site of the author of this manual;

- http://tifaq.calc.org—A "Frequently Asked Questions" page maintained by Ray Kremer; and

- http://www.ticalc.org.

Examples

Here are the details for using the TI-85 for several of the examples from the textbook. Also given are the keystrokes necessary to produce some of the commands shown in the text's examples. In some cases, some suggestions are made for using the calculator more efficiently.

We first consider examples from the text's Appendices, as the calculator techniques they illustrate are useful throughout the text.

Throughout this section, it is assumed that the textbook is available for reference. The problems from the text are not restated here, and there are frequent references to the calculator screens shown in the text.

Appendix A   Example 1   (page 434)   Solving a Linear Equation

Here is a general discussion of how to use the TI-85 to solve (or confirm solutions for) nearly any equation. We will show multiple approaches for solving the equation $\frac{1}{2}x - 6 = \frac{3}{4}x - 9$. (The answer is $x = 12$.) These procedures can be adapted for any equation, including the one from this example, or those found throughout the text.

There are two graphical methods that can be used to confirm this solution. The first is the **intersection** method. To begin, set up the TI-85 to graph the left side of the equation as y1, and the right side as y2. **Note:** Putting the fractions in parentheses ensures no mistakes with order of operations. This is important for the TI-85, since it performs implied multiplication before division; see section 8 of the introduction, page 30.

We are looking for an x value that will make the left and right sides of this equation equal to each other, which corresponds to the x-coordinate of the point of intersection of these two graphs.

Next, select a viewing window which shows the point of intersection; we use $[-15, 15] \times [-10, 10]$ for this example. The TI-85 can automatically locate this point using GRAPH:MATH:ISECT (GRAPH MORE F1 MORE F5). Use ▲, ▼ and ENTER to specify which two functions to use (in this case, the only two being displayed). After pressing ENTER to select the second function, the TI-85 will try to find an intersection of the two graphs. The screens below illustrate these steps.

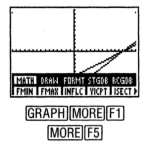
GRAPH MORE F1
MORE F5

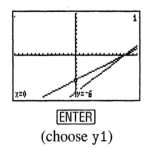
ENTER
(choose y1)

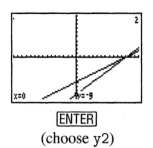
ENTER
(choose y2)

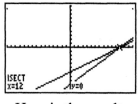
Here is the result.

The x-coordinate of this point of intersection is calculated to 14 digits of accuracy, so if the solution were some less "convenient" number (say, $\sqrt{3}$ or $1/\pi$), we would have an answer that would be accurate enough for nearly any computation.

Note: An approximation for the point of intersection can be found simply by moving the TRACE cursor as near the intersection as possible. The amount of error can be minimized by "zooming in" on the graph. This is the only method available for graphing calculators such as the TI-81.

The second graphical approach is to use the x-**intercept method**, which seeks the x-coordinate of the point where a graph crosses the x-axis. Specifically, we want to know where the graph of y1−y2 crosses the x-axis, where y1 and y2 are as defined above. This is because the equation $\frac{1}{2}x - 6 = \frac{3}{4}x - 9$ can only be true when $\frac{1}{2}x - 6 - \left(\frac{3}{4}x - 9\right) = 0$.

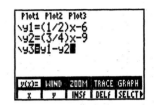

To find this x-intercept, begin by defining y3=y1−y2 on the GRAPH:y(x)= screen. We could do this by re-typing the formulas entered for y1 and y2, but having typed those formulas once, it is more efficient to do this as shown on the right. The simplest way to type "y1" and "y2" is to use the $\boxed{\text{F2}}$ key to produce "y." Note that y1 and y2 have been "de-selected" so that they will not be graphed (see section 12 of the introduction, page 33).

We must first select a viewing window which shows the x-intercept; we again use $[-15, 15] \times [-10, 10]$. The TI-85 can automatically locate this point with the GRAPH:MATH:ROOT ($\boxed{\text{GRAPH}}$$\boxed{\text{MORE}}$$\boxed{\text{F1}}$$\boxed{\text{F3}}$) feature ("root" is a synonym for "x-intercept"). Simply select the function

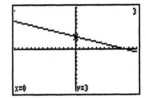 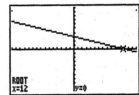

for which a root is sought, and after pressing $\boxed{\text{ENTER}}$, the TI-85 will attempt to find that root. (Provided there is only one root between the bounds, and the function is "well-behaved"—meaning it has some nice properties like continuity—the calculator will find it.) The screens on the right illustrate these steps.

The TI-85 also offers some non-graphical approaches to solving this equation (or confirming a solution): As illustrated on the right, the TI-85's Solver function attempts to find a value of x that makes the given expression equal to 0, given a guess (10, in this case). The solution is stored in the variable x, but as the screen

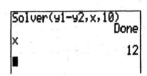

shows, this solution is not automatically displayed. The entry shown use of the fact that y1 and y2 have been defined as the left and right sides of this equation; if that had not been the case, the same results could have been attained by entering (e.g.) Solver((1/2)x-6-((3/4)x-9),x,10). Full details on how to use this function (found in the CATALOG) can be found in the TI-85 manual.

Finally, the TI-85 includes an "interactive solver," accessed with $\boxed{\text{2nd}}$$\boxed{\text{GRAPH}}$. This prompts for the equation to be solved (use $\boxed{\text{ALPHA}}$$\boxed{\text{STO▶}}$ to type the equals sign), then allows the user to enter a guess for the solution (or a range or numbers between which a solution should be sought). To solve the equation, place the cursor on the line beginning with x= and press $\boxed{\text{F5}}$.

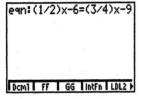

Enter equation

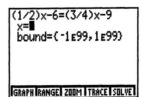

Specify guess, or press $\boxed{\text{F5}}$

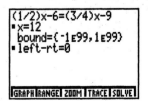

Here is the solution.

The solver can also be used with equations containing more than one variable; simply provide values for all but one variable, then place the cursor on the line containing the variable for which a value is needed and press $\boxed{\text{F2}}$.

Note: In this example, we learned how the TI-85 can be used to support an analytic solution. But the TI-85 and any other graphing calculator also can be used for solving problems when an analytic solution is **not** possible—that is, when one cannot solve an equation "algebraically." This is often the case in many "real-life" applications, and is one of the best arguments for the use of graphing calculators.

Appendix A Example 4 (page 436)	Using the Zero-Factor Property

Appendix A Example 5 (page 437)	Using the Square-Root Property

Appendix A Example 6 (page 437)	Using the Quadratic Formula

The TI-85 can solve quadratic equations (as well as higher-degree polynomial equations) using its built-in polynomial solver, accessed through $\boxed{\text{2nd}}\boxed{\text{PRGM}}$ (POLY). This first prompts the user for the "order" (degree) of the polynomial, meaning the highest power of x. For the quadratic equations, this should be 2.

Pressing $\boxed{\text{ENTER}}$ then brings up the screen on the right, requesting the coefficients of the equation. Note that the top line of the screen contains a reminder that the expression must be equal to 0. The menu at the bottom indicates that $\boxed{\text{F1}}$ will clear the coefficients, while $\boxed{\text{F5}}$ solves the equation. The entries shown here are for Example 6.

Pressing $\boxed{\text{F5}}$ reports the solutions. $\boxed{\text{F1}}$ allows the user to change the coefficient values (that is, it goes back to the previous screen), and $\boxed{\text{F2}}$ provides a way to store the coefficients in a variable (as a list).

Appendix A Example 7 (page 438)	Solving a Linear Inequality

Shown is one way to visualize the solution to inequalities like these. The inequality symbols $>$, $<$, $\geq$, $\leq$ are found in the TEST menu ($\boxed{\text{2nd}}\boxed{\text{2nd}}\boxed{\times}$). The TI-85 responds with 1 when a statement is true and 0 for false statements. This is why the graph of $y1$ appears as it does: For values of x less than 4, the inequality is true (and so $y1$ equals 1), and for $x \geq 4$, the inequality is false (and $y1$ equals 0). Note that this picture does *not* help one determine what happens when $x = 4$.

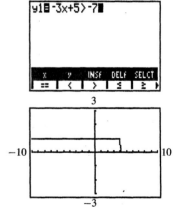

Appendix A   Example 8   (page 439)   Solving a Three-Part Inequality

To see a "picture" of this inequality like the one shown for the previous example requires a little additional work. Setting y1=-2<5+3x≤20 will *not* work correctly. Instead, one must enter

   y1=(-2<5+3x)(5+3x≤20)

which will produce the desired results: A function that equals 1 between $-\frac{7}{3}$ and 5, is equals 0 elsewhere.

Appendix B   Example 2   (page 444)   Using the Midpoint Formula

The TI-85 can do midpoint computations nicely by putting coordinates in parentheses, as shown on the right. (The TI-85 interprets an ordered pair such as $(8,-4)$ as the complex number $8 - 4i$, but since adding two complex numbers means adding their corresponding parts, the computations are done in the correct way to find the midpoint.)

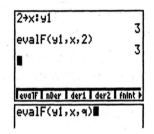

Appendix C   Example 5   (page 455)   Using Function Notation

The TI-85 has a feature that is similar to function notation. Enter the function $(-x^2 + 5x - 3)$ in y1 (or any other function). Return to the home screen, and type either of the expressions on the right. The first simply places 2 in x then computes y1 (which always returns the value based on the current value of x). The second uses the evalF ("evaluate formula") function, found in the CALC menu ([2nd][÷]), which requires three arguments: the expression to be evaluated, the variable, and the value to be used in place of that variable. The result shown on the screen

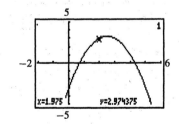

agrees with (a) in the text. [The more natural entry "y1(2)" does not work; it is interpreted as (the value of y1 with the current value of x) times 2.]

   The TI-85 is less helpful if asked to evaluate expressions like those given in Example 5(b). For example, "q→x:y1" or "evalF(y1,x,q)" produces an error message or an unpredictable result (depending on how the variable q is defined), rather than the desired result $-q^2 + 5q - 3$.

To get graphical confirmation, enter the appropriate formula for y1 and graph in any window that includes $x = 2$. Press TRACE ([GRAPH][F4]), then press the [▶] and [◀] keys to change the value of x. It may not be possible to make x exactly equal to 2 in this manner (see section 12 of the introduction, page 33), but rough confirmation that $f(2) = 3$ can be found by observing that y is close to 3 when x is close to 2.

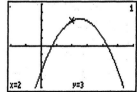

The TI-85 makes it possible to "trace" to any real number value for x between xMin and xMax using the GRAPH:EVAL command ([GRAPH][MORE][MORE][F1]). After choosing this command, simply type a number or expression (like $1/\pi$ or $\sqrt{2}$). Pressing [ENTER] causes the cursor to jump to that x-coordinate.

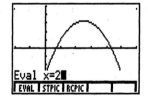

Section 1.1	Example 2	(page 4)	Calculating with Degrees, Minutes, and Seconds

Section 1.1	Example 3	(page 4)	Converting between Decimal Degrees and Degrees, Minutes, and Seconds

See section 9 of this chapter's introduction (page 30) for information about selecting Degree mode. The alternative to putting the calculator in Degree mode is to use the degree symbol following each angle measure.

The MATH:ANGLE menu ([2nd][×][F3]) is shown on the right. It includes the degrees and minutes symbols, and the ▶DMS operator (which causes an angle to be displayed in degrees, minutes, and seconds, rather than as a decimal).

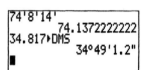

Entering angles on a TI-85, however, is different than on a TI-83 (which was used to create the screens shown in the text.) The degree and seconds symbols (° and ") are not used; the minutes symbol is used for all three positions, as in the screen on the right. (The results of the computations are displayed with the "usual" symbols.)

The screen on the right shows the proper way to enter the angles from Example 3 for computation on the TI-85.

Section 1.2	Example 1	(page 10)	Finding Angle Measures

See page 36 for general information about solving equations using the TI-85. (Of course, one must use knowledge from geometry to obtain the equation in the first place.)

Section 1.3	Example 4	(page 23)	Finding Function Values of Quadrantal Angles

The alternative to putting the calculator in Degree mode is to use the degree symbol ([2nd][×][F3][F1]) following each angle measure; e.g., enter `sin 90°` rather than just `sin 90`. (However, see the comment in the next example.)

Since the cotangent, secant, and cosecant functions are the reciprocals of the tangent, cosine, and sine, they can be entered as (e.g.) `1/sin 90`. (This is illustrated in Figure 31 on page 27 of the text.) Note, though, that this will not properly compute $\cot 90°$, since `1/tan 90` produces a domain error. Entering $\cot x$ as $\cos x / \sin x$ will produce the correct result at $90°$.

One might guess that the other three trigonometric functions are accessed with [2nd] followed by [SIN], [COS], or [TAN] (which produce, e.g., $\sin^{-1}$). This is **not** what these functions do; in this case, the exponent -1 does not mean "reciprocal," but instead indicates that these are inverse functions (which are discussed in Sections 2.3 and 6.1 of the text). The text comments on this distinction at the bottom of page 27.

Section 2.3   Example 1   (page 62)   Finding Function Values with a Calculator

Recall that the TI-85 uses a slightly different format for entering angles in degrees and minutes, as illustrated on the right. This screen also shows how computations like the secant in part (b) can be entered on a single line. These computations were done in Degree mode.

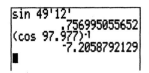

The screen on the right (with computations done in Radian mode) illustrates a somewhat unexpected behavior: Even if an angle is entered in DMS format, the TI-85 assumes that the angle is in radians. In order to remedy this, either put the calculator in Degree mode, or use the degree symbol ([2nd][×][F3][F1]) as was done in the second entry.

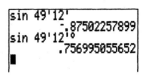

Section 2.3   Example 2   (page 62)   Using an Inverse Trigonometric Function to Find an Angle

The sin⁻¹ ("inverse sine," or "arcsine") function is [2nd][SIN], while cos⁻¹ is [2nd][COS]. The computations shown in Figure 16 in the text were done in Degree mode; the screen on the right shows the result when done in Radian mode. Note that in the second entry, an attempt was made to get the TI-85 to report the result in degrees (by placing the degree symbol at the end of the entry), but this does not have the desired result.

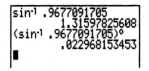

Section 2.4   Example 1   (page 69)   Solving a Right Triangle Given an Angle and a Side

Section 2.4   Example 2   (page 70)   Solving a Right Triangle Given Two Sides

For problems like these, the Ans variable can be used to avoid loss of accuracy from rounding off intermediate results. Shown are computations for Example 1. (Be sure the TI-85 is in Degree mode. Also, recall that the arcsine (or inverse sine) function is [2nd][SIN] —*not* [SIN][2nd][EE].)

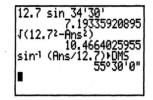

Section 2.5   Example 4   (page 79)   Solving a Problem Involving Angles of Elevation

The TI-85's GRAPH:MATH:ISECT ([GRAPH][MORE][F1][MORE][F5]) feature will automatically locate the intersection of two graphs. This feature was previously illustrated on page 36, but we repeat the description here: Use [▲], [▼] and [ENTER] to specify which two functions to use (in this case, the only two being displayed). After [ENTER] is pressed the second time, the TI-85 will try to find an intersection of the two graphs. The screens below illustrate these steps; the final result is essentially the same as the screen shown in text Figure 32.

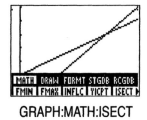

GRAPH:MATH:ISECT

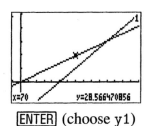

[ENTER] (choose y1)

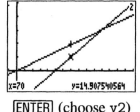

[ENTER] (choose y2)

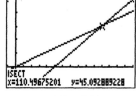

and that's it.

Section 3.1   Example 1   (page 95)   Converting Degrees to Radians

The number π is available as [2nd][∧], and the degree symbol is [2nd][×][F3][F1]. With the calculator in Radian mode (see page 30), entering $45°$ causes the TI-85 to automatically convert to radians.

A useful technique to aid in recognizing when an angle is a multiple of π is to divide the result by π. This approach is illustrated in the screen on the right, showing that $45°$ is $\pi/4$ radians, and $30°$ is $\pi/6$ radians. This screen also makes use of the ▸Frac command from the MATH:MISC menu ([2nd][×][F5][MORE][F1]), which simply means "display the result of this computation as a fraction, if possible." This is a useful enough command that one may wish to put it in the CUSTOM menu (see page 30).

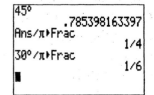

An alternative to using the degree symbol is to store $\pi/180$ in the calculator variable D (see page 30). Then typing, for example, 45D [ENTER] will multiply 45 by $\pi/180$. This approach will work regardless of whether the calculator is in Degree or Radian mode. (A value stored in a variable will remain there until it is replaced by a new value.)

Section 3.1   Example 2   (page 95)   Converting Radians to Degrees

With the TI-85 in Degree mode (see page 30), the radian symbol (a superscripted r), produced with [2nd][×] [F3][F2], will automatically change a radian angle measurement to degrees.

Alternatively, with the value $180/\pi$ stored in the calculator variable R (see page 30), typing $(9\pi/4)$R [ENTER] will convert from radians to degrees regardless of whether the calculator is in Degree or Radian mode. (The same result can be achieved by *dividing by* the calculator variable D as defined in the previous example.)

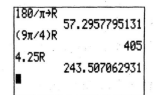

Section 3.3   Example 4   (page 112)   Finding a Number Given Its Circular Function Value

The arccosine (or inverse cosine) function is [2nd][COS] — *not* [COS][2nd][EE]. Likewise, $\tan^{-1}$ is [2nd][TAN].

Section 4.1   Example 1   (page 135)   Graphing $y = a \sin x$

Note that the TI-85 must be in Radian mode in order to produce the desired graph.

The graphs in the text are shown in the "trig viewing window," described on the top of page 135. The TI-85's default trig window (selected with [GRAPH][F3][MORE][F3]) is slightly wider than that shown in the text; it shows x from -2.625π to 2.625π. (If the TI-85 is in Degree mode, xMin and xMax will be ±472.5 instead of $\pm2.625\pi$.) The graph below is shown using the text's trig window, rather than the TI-85's.

The TI-85 does not support varying the thickness of graphs; all graphs appear the same. However, the TRACE ([GRAPH][F4]) feature can be used to determine which graphs correspond to which formulas: On the right, the trace cursor is on graph 2—that is, the graph of $y_2 = \sin x$.

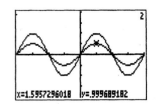

Section 4.2   Example 6   (page 151)   Modeling Temperature with a Sine Function

Given a set of data pairs (x, y), the TI-85 can produce a scatter diagram (like the points shown in Figure 19) and can find various formulas that approximate the relationship between x and y. These formulas are called "regression formulas." The TI-85 can find linear and other polynomial functions, as well as more complex formulas like exponential and logarithmic functions. Unfortunately, the "sine regression" illustrated at the end of this example is not available on the TI-85, but programs may be available to perform this task. See section 13 of this chapter's introduction (page 34) for information about installing and running programs on the TI-85.

Although the TI-85 will not perform a sine regression, we will give the steps for producing a scatter diagram and (for reference) performing other types of regression.

The first step is to enter the data into the TI-85. This is done by pressing STAT F2, producing the screen on the right, requesting which lists should be used to store the values of x and y. The defaults, xStat and yStat, can be selected by pressing ENTER twice.

This brings up the data-entry screen shown on the right. If any data (x and y values) are listed here, it is best to press F5 to clear the old data. Then enter each pair (x, y), starting (in this example) with x₁ = 1, y₁ = 54, etc. (For the scatter diagram shown in the text, enter the temperatures twice—24 pairs of number altogether.)

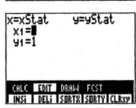

When all data is entered, the screen will look like the one on the right. Note that the last values are x₈ and y₈, so eight pairs have been entered, and the table in the text lists eight pairs. It is a good idea to check the number of entries, and also to proofread the numbers entered.

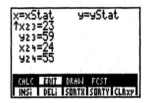

To produce the scatter diagram, set up the viewing window (GRAPH:RANGE) to match the one shown in Figure 19 of the text. Then check that nothing else will be plotted (press GRAPH F1 and make sure that all the equals signs are not highlighted). Finally, press STAT F3 F2 (STAT:DRAW:SCAT) to produce the graph. The results are shown on the right; note that the TI-85 just draws single dots at each point, which can make them rather difficult to see.

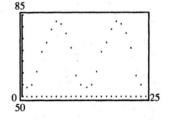

To find a regression formula, press STAT F1 (CALC), at which point the calculator once again asks which lists contain the data (as shown on the right). As before, press ENTER twice to accept the defaults.

The bottom of the screen now lists the various options for the type of calculation to be done. $\boxed{\text{F2}}$ is LINR, short for linear regression; also available are LNR (logarithmic regression), exponential regression, power regression, and (after pressing $\boxed{\text{MORE}}$) polynomial regressions for degrees two through five (that is, fitting second-, third-, fourth-, and fifth-degree polynomials).

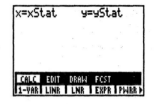

Section 4.3   Example 1   (page 158)   Graphing $y = a\sec bx$

Section 4.3   Example 2   (page 159)   Graphing $y = a\csc(x - d)$

See section 9 of the introduction (page 30) for information about DrawLine versus Dot (DrawDot) mode. The function in Example 1 can be entered as y1=2cos (x/2)⁻¹ (or as shown in the text). The function in Example 2 can be entered as ẏ1=(3/2)(1/sin (x-π/2)) or y1=3/(2sin (x-π/2)). A reminder: sin⁻¹ ($\boxed{\text{2nd}}\boxed{\text{SIN}}$) is *not* the cosecant function.

Section 5.1   Example 3   (page 185)   Rewriting an Expression in Terms of Sine and Cosine

The screen on the right shows how these expressions are entered on the "y equals" screen.

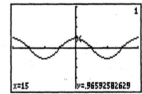

Section 5.3   Example 1   (page 198)   Finding Exact Cosine Function Values

---

Section 5.4   Example 1   (page 206)   Finding Exact Sine and Tangent Function Values

The TI-85 can graphically and numerically support exact value computations such as $\cos 15° = \frac{\sqrt{6}+\sqrt{2}}{4}$. Starting with a graph of y1=cos x, the TI-85 makes it possible to evaluate y1 (or any function) for any value of x between xMin and xMax using the TI-85's GRAPH:EVAL ($\boxed{\text{GRAPH}}$

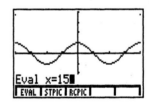

$\boxed{\text{MORE}}\boxed{\text{MORE}}\boxed{\text{F1}}$) option. GRAPH:EVAL prompts for a value of x, which can be either a number like 15 or expression (like 1/π or √2). Pressing $\boxed{\text{ENTER}}$ causes the cursor to jump to that x-coordinate.

Note that Example 1 in Section 5.6 (page 222) shows that $\cos 15°$ can also be written as $\frac{\sqrt{2+\sqrt{3}}}{2}$.

Section 6.1   Example 1   (page 238)   Finding Inverse Sine Values

Section 6.1   Example 2   (page 239)   Finding Inverse Cosine Values

Of course, it is not necessary to graph $y = \sin^{-1} x$ to find these values; one can simply enter, e.g., `sin⁻¹ (1/2)` on the home screen. The first entry of the screen on the right shows what happens when the calculator is in Degree mode; note that the result is not in $[-\pi/2, \pi/2]$. With the calculator in Radian mode, results similar to those in the text are found, and the method employed on page 42 of this manual (in the discussion of text Example 1 from Section 3.1) confirms that these values are $\pi/6$ and $3\pi/4$.

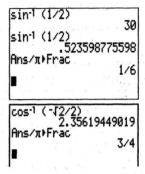

Note that the TI-85 does *not* give an error for the input `sin⁻¹ -2`, but instead gives a complex result: $-\pi/2 + i\ln(2 + \sqrt{3})$. The reason that this is technically a correct result is beyond the scope of the textbook; TI-85 owners should recognize that a complex answer is not appropriate for this problem, and so should ignore this result.

Section 6.1   Example 4   (page 243)   Finding Inverse Function Values with a Calculator

Note that the answer given for (c), 109.499054°, overrepresents the accuracy of that value. A typical rule for doing computations involving decimal values (like -0.3541) is to report only as many digits in the result as were present in the original number—in this case, four. This means the reported answer should be "about 109.5°," and in fact, any angle θ between about 109.496° and 109.501° has a cotangent which rounds to -0.3541. (See the discussion of significant digits on page 68 of the text.)

Section 6.2   Example 6   (page 253)   Describing a Musical Tone from a Graph

Section 6.3   Example 5   (page 259)   Analyzing Pressures of Upper Harmonics

Note that the calculator screens shown in Figures 23–26 illustrate the importance of choosing a "good" viewing window. If we choose the wrong vertical scale (`yMin` and `yMax`), we might not be able to see the graph at all—it might be squashed against the *x*-axis. If we make the window too wide—that is, if `xMax` minus `xMin` is too large—we might see the "wrong" picture, like the one on the right (for Example 5 from Section 6.3): We seem to see six periods in

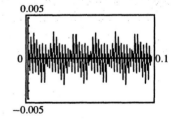

this view, when in fact there are 44. Each of the six "pseudo-cycles" is made of parts of seven or eight full periods.

This observation—that a periodic function, viewed at fixed intervals, can appear to be a *different* periodic function—is the same effect that causes wagon wheels to appear to run backwards in old movies.

Section 7.4   Example 1   (page 307)   Finding Magnitude and Direction Angle

Section 7.4   Example 2   (page 307)   Finding Horizontal and Vertical Components

The TI-85 does not have the conversion functions shown in Figures 26 and 28, but can do the desired conversions in a manner that is perhaps even more convenient. The TI-85 recognizes vectors entered in either of two formats:

[*horizontal component , vertical component*]   or   [*magnitude ∠ angle*]

(The square brackets are [2nd][(] and [2nd][)], and "∠" is [2nd][,].) Regardless of how the vector is entered, the TI-85 displays it according to the <u>RectV CylV SphereV</u> mode setting (see page 30); specifically, it displays the vector in component form in RectV mode, and in magnitude/angle form for either of the other two modes.

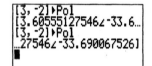

For Example 1, then, either put the TI-85 in CylV or SphereV mode, or (perhaps more conveniently), use the ▸Pol command ([2nd][8][F4][F3]), as the screen on the right illustrates. This causes a vector to be displayed in magnitude/direction angle format regardless of the mode setting. On the fourth line, we see the results of pressing [▸] to see all the digits of the angle. (With the TI-85 in Degree mode, the angle returned is in degrees.)

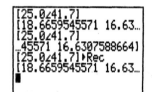

For Example 2, both the horizontal and vertical components can be found at once, as the screen on the right shows. (The vertical component is too long to fit on the screen, but the other digits can be seen by pressing [▸]. Note that the screen shown in the text was created with the calculator set to display only one digit after the decimal.) These computations were done with the TI-85 in Degree mode, so it was not necessary to include the degree symbol on the angle. The third entry shown uses the ▸Rec command, which forces the vector to be displayed in component form regardless of the mode setting. (It was unnecessary in this case, because the TI-85 with in RectV mode.)

Finally, observe that this conversion could be done by typing 25cos(41.7) and 25sin(41.7); which approach to use is a matter of personal preference.

Section 7.4   Example 3   (page 308)   Writing Vectors in the Form ⟨a, b⟩

Note how easily these conversions are done in the TI-85's polar vector format. (The TI-85 was in Degree mode.)

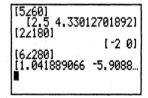

Section 7.4   Example 5   (page 309)   Performing Vector Operations

---

Section 7.4   Example 6   (page 310)   Finding Dot Products

Using the TI-85's vector notation, the vectors can be stored in calculator variables
(see page 30) which can then be used to do the desired operations. Shown are the
commands to compute 5(c): $4\mathbf{u} - 3\mathbf{v}$, and 6(a): $\langle 2, 3 \rangle \cdot \langle 4, -1 \rangle$ (the dot command
is 2nd 8 F3 F4).

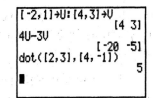

Section 8.1   Example 1   (page 333)   Writing $\sqrt{-a}$ as $i\sqrt{a}$

Section 8.1   Example 4   (page 334)   Finding Products and Quotients Involving Negative Radicands

The TI-85 does these computations almost as shown in the text, except that it
reports the results differently. The screen on the right illustrates the output from
the computations for parts (b) and (d) of Example 4. The TI-85 uses the format
(a, b) for the complex number $a + bi$; whenever the TI-85 does a computation
involving square roots of negative numbers, it reports the results in this format *even if the result is real*. Thus
$(-7.74596669241, 0)$ represents the real number $-7.74596669241 \approx -2\sqrt{15}$, and $(0, 1.41421356237)$
is $i\sqrt{2}$.

The calculator always reports complex *results* in this format, or in polar format,
which is similar to the text's trigonometric format. However, it can recognize
input typed in using "$a + bi$" format by first defining a variable i as the complex
number $(0, 1) = 0 + 1i$. Shown on the right is a computation using this definition.
One could also use I (uppercase), which can be typed with one less keystroke, but i will be used in these
examples.

Section 8.1   Example 2   (page 333)   Solving Quadratic Equations for Complex Solutions

---

Section 8.1   Example 3   (page 334)   Solving a Quadratic Equation for Complex Solutions

The TI-85's POLY feature can find complex solutions as
well as real ones. In the screens on the right, the TI-85
was set to display four digits after the decimal.

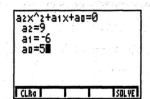

| Section 8.1 | Example 5 | (page 335) | Adding and Subtracting Complex Numbers |

| Section 8.1 | Example 6 | (page 336) | Multiplying Complex Numbers |

| Section 8.1 | Example 7 | (page 337) | Simplifying Powers of *i* |

| Section 8.1 | Example 8 | (page 338) | Dividing Complex Numbers |

Using the variable i defined to be (0, 1)—as was described in the previous example—computations like these can be done in a manner similar to those illustrated in the text's calculator screens. The screens on the right show (in the TI-85's format) the same results as those in the text accompanying Examples 6 and 8.

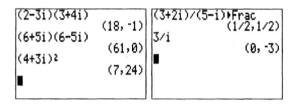

| Section 8.2 | Example 2 | (page 342) | Converting from Trigonometric Form to Rectangular Form |

Complex numbers can be entered in trigonometric form in a manner similar to the polar format for vectors (see page 46); the screen on the right illustrates how to quickly convert from the TI-85's trigonometric form to its rectangular form (assuming the calculator is in RectC mode; see page 30). "∠" is [2nd][,].

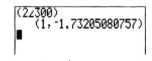

| Section 8.2 | Example 3 | (page 343) | Converting from Rectangular Form to Trigonometric Form |

Although the TI-85 does not have the conversion functions shown in the text, the same thing can be accomplished with the abs and angle commands, found in the TI-85's CPLX menu ([2nd][9]), and shown on page 347 of the text. On the screen on the right, note that if one makes use of the variable i defined as (0, 1), the angle and abs functions require parentheses; the third output is incorrect because it first computes the "angle" of the real number $-\sqrt{3}$ (which is 0, although it really should be π) and then adds this to i.

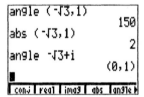

More convenient is the TI-85's ▸Pol command, found by pressing [MORE] in the CPLX menu. (This command is also found in the VECTR:OPS menu.) This gives both the magnitude and angle at the same time.

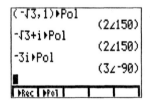

| Section 8.2 | Example 4 | (page 344) | Converting Between Trigonometric and Rectangular Forms Using Calculator Approximations |

Aside from entering the expression in (a) as it is shown in the text (using the cos and sin functions), one can use the TI-85's polar format $(r \angle \theta)$, but be sure to either put the TI-85 in the correct mode (Degree or Radian) or use the degrees or radians symbol from the MATH:ANGLE menu. Both computations on the right were done in Radian mode, and the first result is incorrect.

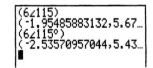

Section 8.3	Example 1	(page 348)	Using the Product Theorem

Section 8.3	Example 2	(page 349)	Using the Quotient Theorem

Section 8.3	Example 3	(page 349)	Using the Product and Quotient Theorems with a Calculator

The TI-85's polar complex format makes these computations very easy to enter.

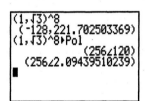

Section 8.4	Example 1	(page 353)	Finding a Power of a Complex Number

The screen on the right shows several options for computing $(1 + i\sqrt{3})^8$ with the TI-85. The first is fairly straightforward, but the reported result shows the complex part of the answer $(128 i\sqrt{3})$ given in decimal form. For an "exact" answer, the TI-85's polar format can be used. The TI-85 was in Degree mode for the first of the two polar-format answers, and in Radian mode for the second (the angle given in the second answer is $2\pi/3$).

Section 8.4	Example 4	(page 356)	Solving an Equation by Finding Complex Roots

The TI-85's POLY polynomial solver (see page 38) can find complex solutions as well as real ones, and can be used to see the pattern of these roots, if the calculator is in PolarC and Degree modes. The two negative angles shown in the second screen are coterminal with the angles 216° and 288° given in the text.

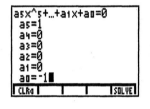

Section 8.5	Example 3	(page 361)	Examining Polar and Rectangular Equations of Lines and Circles

Section 8.5	Example 4	(page 363)	Graphing a Polar Equation (Cardioid)

Section 8.5	Example 5	(page 363)	Graphing a Polar Equation (Rose)

Section 8.5	Example 6	(page 364)	Graphing a Polar Equation (Lemniscate)

To produce these polar graphs, the TI-85 should be set to Degree and Polar modes (see the screen on the right). In this mode, GRAPH F1 allows entry of polar equations (r as a function of θ). One could also use Radian mode, adjusting the values of θMin, θMax, and θStep accordingly (e.g., use 0, 2π, and $\pi/30$ instead of 0, 360, and 5).

For the cardioid, rose, and lemniscate, the window settings shown in the text show these graphs on "square" windows (see section 11 of the introduction, page 32), so one can see how their proportions compare to those of a circle. Note, however, that on the TI-85, these window are not square; the ZOOM:ZSQR option can be used to adjust the window dimensions to make it square.

For the cardioid, the value of θStep does not need to be 5, although that choice works well for this graph. Too large a choice of θStep produces a graph with lots of sharp "corners," like the one shown on the right (drawn with θStep=30). Setting θStep too small, on the other hand, produces a smooth graph, but it is drawn very slowly. Sometimes it may be necessary to try different values of θStep to choose a good one.

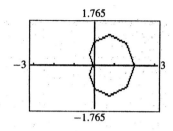

The lemniscate can be drawn by setting θMin=0 and θMax=180, or θMin=-45 and θMax=45. In fact, with θ ranging from −45 to 225, the graph of r1=√cos (2θ) (alone) will produce the entire lemniscate. (θStep should be about 5.)

The rose can be produced by setting θMin=0 and θMax=360, or using any 360°-range of θ values (with θStep about 5).

Section 8.5   Example 7   (page 365)   Graphing a Polar Equation (Spiral of Archimedes)

To produce this graph on the viewing window shown in the text, the TI-85 must be in Radian mode. (In Degree mode, it produces the same shape, but magnified by a factor of $180/\pi$ — meaning that the viewing window needs to be larger by that same factor.)

Section 8.6   Example 1   (page 371)   Graphing a Plane Curve Defined Parametrically

---

Place the TI-85 in Parametric mode, as the screen on the right shows. In this mode, GRAPH F1 allows entry of pairs of parametric equations (x and y as functions of t). No graph is produced unless both functions in the pair are entered and selected (that is, both equals signs are highlighted).

The value of tStep does not need to be 0.05, although that choice works well for this graph. Too large a choice of tStep produces a less-smooth graph, like the one shown on the right (drawn with tStep=1). Setting tStep too small, on the other hand, produces a smooth graph, but it is drawn very slowly. Sometimes it may be necessary to try different values of tStep to choose a good one.

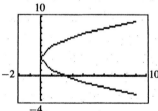

Section 8.6   Example 3   (page 372)   Graphing a Plane Curve Defined Parametrically

---

This curve can be graphed in Degree mode with tMin=0 and tMax=360, or in Radian mode with tMax=2π. In order to see the proportions of this ellipse, it might be good to graph it on a square window. This can be done most easily with the ZOOM:ZSQR option (GRAPH F3 MORE F2). On a TI-85, initially with the window settings shown in the text, this would result in the window [−6.8, 6.8] × [−4, 4].

Section 8.6   Example 5   (page 373)   Graphing a Cycloid

The TI-85 *must* be in Radian mode in order to produce this graph.

Section 8.6   Example 6   (page 374)   Simulating Motion with Parametric Equations

Section 8.6   Example 8   (page 375)   Analyzing the Path of a Projectile

Parametric mode is particularly nice for analyzing motion, because one can picture the motion by watching the calculator create the graph, or by using TRACE ([GRAPH][F4]) and watching the motion of the trace cursor. (When tracing in parametric mode, the [▶] and [◀] keys increase and decrease the value of t, and the trace cursor shows the location (x, y) at time t.) Figure 40 illustrates tracing on the projectile path in Example 8. Note that the value of t changes by ±tStep each time [▶] or [◀] is pressed, so obviously the choice of tStep affects which points can be traced.

Section 9.1   Example 4   (page 393)   Using a Property of Exponents to Solve an Equation

Below the calculator screen shown in the text, the caption refers to "the x-intercept method of solution." This and other methods for solving equations were described beginning on page 36 of this manual.

Section 9.1   Example 11   (page 397)   Using Data to Model Exponential Growth

The scatter diagram in Figure 10(a) and the "exponential regression" in Figure 12 can be reproduced by adapting the procedures described on page 43 of this manual. The corresponding command on the TI-85 is EXPR ([STAT][F1][ENTER][ENTER][F4]).

The two calculator screens in Figure 11 use the "intersection method" of solving equations; see page 36 for a description.

Section 9.3   Example 1   (page 414)   Finding pH

For (a), the text shows -log(2.5*10^(-4)), but this could also be entered as shown on the first line of the screen on the right, since "E" (produced with [EE]) and "*10^" are nearly equivalent. The two are not completely interchangeable, however; in particular, in part (b), "10^" **cannot** be replaced with "E", because "E" is only valid when followed by an *integer*. That is, E-7 produces the same result as 10^-7, but the last line shown on the screen produces a syntax error.

(Incidentally, "10^" is [2nd][LOG], but [1][0][^] produces the same results.)

Introduction

The information in this section is essentially a summary of material that can be found in the TI-86 manual. Consult that manual for more details.

1   Power

To power up the calculator, simply press the ON key. This should bring up the "home screen"—a flashing block cursor, and possibly the results of any previous computations that might have been done.

If the home screen does not appear, one may need to adjust the contrast (see the next section).

To turn the calculator off, press 2nd ON (note that the "second function" of ON—written in yellow type above the key—is "OFF"). The calculator will automatically shut off if no keys are pressed for several minutes.

2   Adjusting screen contrast

If the screen is too dark (all black), decrease the contrast by pressing 2nd then pressing and holding ▼. If the screen is too light, increase the contrast by pressing 2nd and then press and hold ▲.

As one adjusts the contrast, the numbers 1 through 9 will appear in the upper right corner of the screen. If the contrast setting reaches 8 or 9, or if the screen never becomes dark enough to see, the batteries should be replaced.

3   Replacing batteries

To replace the four AAA batteries, first turn the calculator off (2nd ON), then remove the back cover, remove and replace each battery, replace the back cover, then turn the calculator on again. (After replacing batteries, one may need to adjust the contrast down as described above.)

4   Basic operations

Simple computations are entered in essentially the same way they would be written. For example, to compute $2 + 17 \times 5$, press $\boxed{2}\boxed{+}\boxed{1}\boxed{7}\boxed{\times}\boxed{5}\boxed{\text{ENTER}}$ (the $\boxed{\text{ENTER}}$ key tells the calculator to act on what has been typed). Standard order of operations (including parentheses) is followed.

```
2+17*5
              87
■
```

The result of the most recently entered expression is stored in Ans, which is typed by pressing $\boxed{\text{2nd}}\boxed{(-)}$ (the word "ANS" appears in yellow above this key). For example, $\boxed{5}\boxed{+}\boxed{\text{2nd}}\boxed{(-)}\boxed{\text{ENTER}}$ will add 5 to the result of the previous computation.

```
2+17*5
              87
5+Ans
              92
■
```

After pressing $\boxed{\text{ENTER}}$, the TI-86 automatically produces Ans if the first key pressed is one which requires a number before it; the most common of these are $\boxed{+}$, $\boxed{-}$, $\boxed{\times}$, $\boxed{\div}$, $\boxed{\wedge}$, $\boxed{x^2}$, and $\boxed{\text{STO}\blacktriangleright}$. For example, $\boxed{+}\boxed{5}\boxed{\text{ENTER}}$ would accomplish the same thing as the keystrokes above (that is, it adds 5 to the previous result).

```
2+17*5
              87
5+Ans
              92
Ans+5
              97
■
```

Pressing $\boxed{\text{ENTER}}$ by itself evaluates the previously typed expression again. This can be especially useful in conjunction with Ans. The screen on the right shows the result of pressing $\boxed{\text{ENTER}}$ a second time.

```
2+17*5
              87
5+Ans
              92
Ans+5
              97
             102
■
```

Several expressions can be evaluated together by separating them with colons ($\boxed{\text{2nd}}\boxed{\cdot}$). When $\boxed{\text{ENTER}}$ is pressed, the result of the *last* computation is displayed. The screen shown illustrates the computation $2(5+1)^2$.

```
3+2
               5
Ans+1:Ans²:2 Ans
              72
■
```

5   Cursors

When typing, the appearance of the cursor indicates the behavior of the next keypress. When the standard cursor (a flashing solid block, ■) is visible, the next keypress will produce its standard action—that is, the command or character printed on the key itself.

If $\boxed{\text{2nd}}\boxed{\text{DEL}}$ is pressed, the TI-86 is placed in INSERT mode and the standard cursor will appear as a flashing underscore. If the arrow keys ($\boxed{\blacktriangle}$, $\boxed{\blacktriangledown}$, $\boxed{\blacktriangleright}$, $\boxed{\blacktriangleleft}$) are used to move the cursor around within the expression, and the TI-86 is placed in INSERT mode, subsequent characters and commands will be inserted in the line at the cursor's position. When the cursor appears as a block, the TI-86 is in DELETE (or OVERWRITE) mode, and subsequent keypresses will replace the character(s) at the cursor's position. (When the cursor is at the end of the expression, this is irrelevant.)

The TI-86 will return to DELETE mode when any arrow key is pressed. It can also be returned to DELETE mode by pressing $\boxed{\text{2nd}}\boxed{\text{DEL}}$ a second time.

Pressing $\boxed{\text{2nd}}$ causes an arrow to appear in the cursor: ▯ (or an underscored arrow). The next keypress will produce its "second function"—the command or character printed in yellow above the key. (The cursor will then return to "standard.") If $\boxed{\text{2nd}}$ is pressed by mistake, pressing it a second time will return the cursor to standard.

Pressing ALPHA places the letter "A" in the cursor: **A** (or an underscored "A"). The next keypress will produce the letter or other character printed in blue above that key (if any), and the cursor will then return to standard. Pressing 2nd ALPHA puts the calculator in lowercase ALPHA mode, changing the cursor to **a** and producing the lowercase version of a letter. Pressing ALPHA twice (or 2nd ALPHA ALPHA) "locks" the TI-86 in ALPHA (or lowercase ALPHA) mode, so that all of the following keypresses will produce characters until ALPHA is pressed again.

6   Accessing previous entries ("deep recall")

By repeatedly pressing 2nd ENTER ("ENTRY"), previously typed expressions can be retrieved for editing and re-evaluation. Pressing 2nd ENTER once recalls the most recent entry; pressing 2nd ENTER again brings up the second most recent, etc. The number of previous entries thus displayed varies with the length of each expression (the TI-86 allocates 128 bytes to store previous expressions).

7   Menus

Keys such as TABLE, GRAPH and 2nd 7 (MATRX) bring up a menu line at the bottom of the screen with a variety of options. These options can be selected by pressing one of the function keys (F1, F2, . . . , F5). If the menu ends with a small triangle ("▸"), it means that more options are available in this menu, which can viewed by pressing MORE. Shown is the menu produced by pressing 2nd × (MATH).

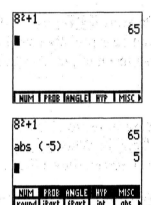

This screen shows the result of pressing F1 (the "NUM" option, which lists a variety of numerical functions). Note that the MATH menu still appears (with NUM highlighted) and the bottom line now lists the functions available in this sub-menu—including, for example, the absolute value function (abs), which is accessed by pressing F5. The command line abs (-5) was typed by pressing F5 ((-) 5) ENTER.

This manual will use (e.g.) MATH:NUM to indicate commands accessed through menus like this. Sometimes the keypresses will be included as well; for this example, it would be 2nd × F1 F5.

The various commands in these menus are too numerous to be listed here. They will be mentioned as needed in the examples.

One last comment is worthwhile, however. Some functions that may be used frequently are buried several levels deep in the menus, and may take many keystrokes to access. Worse, the location of the function might be forgotten (is it MATH:NUM or MATH:MISC?), necessitating a search through the menus. It is useful to remember three things:

- Any command can be typed one letter at a time, in either upper- or lowercase; e.g., ALPHA ALPHA LOG SIN 6 (-) will type the letters "ABS ", which has the same effect as 2nd × F1 F5.

- Any command can be found in the CATALOG menu ([2nd][CUSTOM][F1]). Since the commands appear in alphabetical order, it may take some time to locate the desired function. Pressing any letter key brings up commands starting with that letter (it is not necessary to press [ALPHA] first); e.g., pressing [LOG] brings up the list on the right, while pressing [,] shows commands starting with "P."

- Frequently used commands can be placed in the CUSTOM menu, and will then be available simply by pressing [CUSTOM]. To do this, scroll through the CATALOG to find the desired function, then press [F3] (CUSTM) followed by one of [F1]–[F5] to place that command in the CUSTOM menu. In the screen shown, [F1] was pressed, so that pressing [CUSTOM][F1] will type "Solver(." The commands in the MATH:ANGLE menu ([2nd][×][F3]), used frequently for problems in this text, could be made more accessible by placing them in this menu.

8   Variables

The uppercase letters A through Z, as well as some (but not all) lowercase letters, and also sequences of letters (like "High" or "count") can be used as variables (or "memory") to store numerical values. To store a value, type the number (or an expression) followed by [STO▸], then a letter or letters (note that the TI-86 automatically goes into ALPHA mode when [STO▸] is pressed), then [ENTER]. That variable name can then be used in the same way as a number, as demonstrated at right.

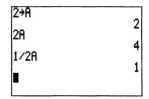

Note: The TI-86 interprets 2A as "2 times A"—the "∗" symbol is not required (this is consistent with how we interpret mathematical notation). As for order of operations, this kind of multiplication is treated the same as "∗" multiplication.

9   Setting the modes

By pressing [2nd][MORE] (MODE), one can change many aspects of how the calculator behaves. For most of the examples in this manual, the "default" settings should be used; that is, the MODE screen should be as shown on the right. Each of the options is described below; consult the TI-86 manual for more details. Changes in the settings are made using the arrows keys and [ENTER].

The Normal Sci Eng setting specifies how numbers should be displayed. The screen on the right shows the number 12345 displayed in Normal mode (which displays numbers in the range ±999, 999, 999, 999 with no exponents), Sci mode (which displays all numbers in scientific notation), and Eng mode (which uses only exponents that are multiples of 3). Note: "E" is short for "times 10 to the power," so $1.2345\text{E}4 = 1.2345 \times 10^4 = 1.2345 \times 10000 = 12345$.

The <u>Float 012345678901</u> setting specifies how many places after the decimal should be displayed (the 0 and 1 at the end mean 10 and 11 decimal places). The default, `Float`, means that the TI-86 should display all non-zero digits (up to a maximum of 12).

<u>Radian Degree</u> indicates whether angle measurements should be assumed to be in radians or degrees. (A right angle measures $\frac{\pi}{2}$ radians, which is equivalent to $90°$.) Most of the examples in the text include reminders to set the calculator in the appropriate mode, in cases where this is important.

<u>RectC PolarC</u> specifies whether complex numbers should be displayed in rectangular or polar format. These two formats are essentially the same as the two used by the textbook. **Note:** The text prefers the term "trigonometric format" rather than "polar format." More information about complex number formats can be found beginning on page 75 of this manual.

<u>Func Pol Param DifEq</u> specifies whether formulas to be graphed are functions (y as a function of x), polar equations (r as a function of θ), parametric equations (x and y as functions of t), or differential equations ($Q'(t)$ as a function of Q and t). The text accompanying this manual uses the first three of these settings.

The <u>RectV CylV SphereV</u> setting indicates the default display format for vectors (see page 74 of this manual).

The other two mode settings deal with issues that are beyond the scope of the textbook, and are not discussed here.

A group of settings related to the graph screen are found by pressing GRAPH MORE F3 (GRAPH:FORMT). The default settings are shown in the screen on the right, and are generally the best choices for most examples in this book (although the last setting could go either way).

<u>RectGC PolarGC</u> specifies whether graph coordinates should be displayed in rectangular (x, y) or polar (r, θ) format. Note that this choice is independent of the `Func Pol Param DifEq` mode setting.

The <u>CoordOn CoordOff</u> setting determines whether or not graph coordinates should be displayed.

When plotting a graph, the <u>DrawLine DrawDot</u> setting tells the TI-86 whether or not to connect the individually plotted points. <u>SeqG SimulG</u> specifies whether individual expressions should be graphed one at a time (sequentially), or all at once (simultaneously).

<u>GridOff GridOn</u> specifies whether or not to display a grid of dots on the graph screen, while <u>AxesOn AxesOff</u> and <u>LabelOff LabelOn</u> do the same thing for the axes and labels (y and x) on the axes.

10   Setting the graph window

Pressing GRAPH F2 brings up the WINDOW settings. The exact contents of the WINDOW menu vary depending on whether the calculator is in function, polar, parametric, or DifEq mode; below are four examples showing this menu in each of these modes.

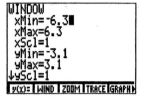
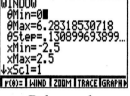
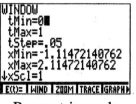

    Function mode   Polar mode   Parametric mode   DifEq mode

All these menus include the values xMin, xMax, xScl, yMin, yMax, and yScl. When GRAPH F5 (GRAPH) is pressed, the TI-86 will show a portion of the Cartesian (x-y) plane determined by these values. In function mode, this menu also includes xRes, the behavior of which is described in section 12 of this manual (page 60). The other settings in this screen allow specification of the smallest, largest, and step values of θ (for polar mode) or t (for parametric mode), or initial conditions for the differential equation.

With settings as in the "Function mode" screen shown above, the TI-86 would display the screen at right: x values from -6.3 to 6.3 (that is, from xMin to xMax), and y values between -3.1 to 3.1 (yMin to yMax). Since xScl $=$ yScl $= 1$, the TI-86 places tick marks on both axes every 1 unit; thus the x-axis ticks are at $-6, -5, \ldots, 5$, and 6, and the y-axis ticks fall on the integers from -3 to 3. This window is called the "decimal" window, and is most quickly set by pressing GRAPH F3 (ZOOM) MORE F4 (ZDECM).

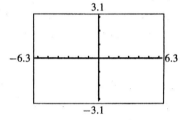

Note: If the graph screen has a menu on the bottom (like that shown on the right), possibly obscuring some important part of the graph, it can be removed by pressing CLEAR. The menu can be restored later by pressing EXIT.

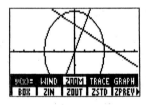

Below are four more sets of window settings, and the graph screens they produce. Note that the first graph on the left has tick marks every 10 units on both axes. The second window is called the "standard" viewing window, and is most quickly set by pressing GRAPH F3 (ZOOM) F4 (ZSTD). The setting yScl $= 0$ in the final graph means that no tick marks are placed on the y-axis.

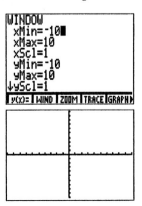
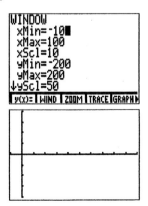

11   The graph screen

The TI-86 screen is made up of an array of rectangular dots (pixels) with 63 rows and 127 columns. All the pixels in the leftmost column have x-coordinate xMin, while those in the rightmost column have x-coordinate xMax. The x-coordinate changes steadily across the screen from left to right, which means that the coordinate for the nth column (counting the leftmost column as column 0) must be $\text{xMin} + n\Delta x$, where $\Delta x = (\text{xMax} - \text{xMin})/126$. Similarly, the nth row of the screen (counting up from the bottom row, which is row 0) has y-coordinate $\text{yMin} + n\Delta y$, where $\Delta y = (\text{yMax} - \text{yMin})/62$.

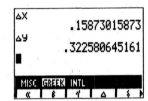

It is not necessary to memorize the formulas for Δx and Δy. Should they be needed, they can be determined by pressing GRAPH F5 and then the arrow keys. When pressing ▶ or ◀ successively, the displayed x-coordinate changes by Δx; meanwhile, when pressing ▲ or ▼, the y-coordinate changes by Δy. Alternatively, the values can be found by typing "Δx" and "Δy" on the home screen; this is most easily done by pressing 2nd 0 F2 F4 to access the CHAR:GREEK menu and type the "Δ" character, then typing lowercase x or y. This produces results like those shown on the right; the CHAR:GREEK menu remains on the bottom of the screen.

In the decimal window $\text{xMin} = -6.3$, $\text{xMax} = 6.3$, $\text{yMin} = -3.1$, $\text{yMax} = 3.1$, note that $\Delta x = 0.1$ and $\Delta y = 0.1$. Thus, the individual pixels on the screen represent x-coordinates $-6.3, -6.2, -6.1, \ldots, 6.1,$ $6.2, 6.3$ and y-coordinates $-3.1, -3, -2.9, \ldots, 2.9, 3, 3.1$. This is where the decimal window gets its name.

It happens that the pixels on the TI-86 screen are about 1.2 times taller than they are wide, so if $\Delta y/\Delta x$ is approximately 1.2 (the exact value is $1.19565\ldots$), the window will be a "square" window (meaning that the scales on the x- and y-axes are equal). For example, the decimal window (with $\Delta y/\Delta x = 1$) is not square, so that one unit on the x-axis is not the same length as one unit on the y-axis. (Specifically, one y-axis unit is about 20% longer than one x-axis unit.)

Any window can be made square be pressing GRAPH F3 (ZOOM) MORE F2 (ZSQR). To see the effect of a square window, observe the two pairs of graphs below. In each pair, the first graph is on the standard window, and the second is on a square window (after choosing ZOOM:ZSQR). The first pair shows the lines $y = 2x - 3$ and $y = 3 - \frac{1}{2}x$; note that on the square window, these lines look perpendicular (as they should). The second pair shows a circle centered at the origin with a radius of 8. On the standard window, this looks like an oval since the screen is wider than it is tall. (The reason for the gaps in the circle will be addressed in the next section.)

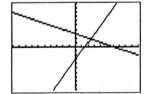 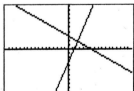 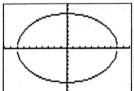 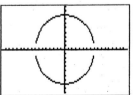

12   Graphing a function

This introductory section only addresses creating graphs in function mode. Procedures for creating parametric and polar graphs are very similar; they are covered beginning on page 77 of this manual, in material related to Chapter 8 of the text.

To see the graph of $y = 2x - 3$, begin by entering the formula into the calculator. This is done by pressing [GRAPH][F1] to access the "y equals" screen of the calculator. Enter the formula as y1 (or any other yn); note that the letter x can be typed by pressing [F1] or [x-VAR] (as well as [2nd][ALPHA][+]). If another y variable has a formula, position the cursor on that line and press either [F4] (DELf—to delete the function) or [F5] (SELCT). The latter has the effect of toggling the "highlighting" for the equals sign "=" for that line (an "unhighlighted" equals sign tells the TI-86 not to graph that formula). In the screen on the right, only y1 will be graphed.

The next step is to choose a viewing window. See the previous section for more details on this. This example uses the standard window ([GRAPH][F3][F4]).

If the graph has not been displayed, press [GRAPH][F5], and the line should be drawn. In order to produce this graph, the TI-86 considers 127 values of x, ranging from xMin to xMax in steps of Δx (assuming that xRes = 1; see below for other possibilities). For each value of x, it computes the corresponding value of y, then plots that point (x, y) and (if the calculator is in Connected [DrawLine] mode) draws a line between this point and the previous one.

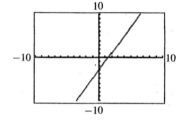

If xRes is set to 2, the TI-86 will only compute y for every other x value; that is, it uses a step size of $2\Delta x$. Similarly, if xRes is 3, the step size will be $3\Delta x$, and so on. Setting xRes higher causes graphs to appear faster (since fewer points are plotted), but for some functions, the graph may look "choppy" if xRes is too large, since detail is sacrificed for speed.

Note: If the line does not appear, or the TI-86 reports an error, double-check all the previous steps. Also, check the mode settings (discussed in section 9, page 56).

Once the graph is visible, the window can be changed using [F2] (WINDOW) or [F3] (ZOOM). Pressing [F4] (TRACE) brings up the "trace cursor," and displays the x- and y-coordinates for various points on the line as the [◄] and [►] keys are pressed. Tracing beyond the left or right columns causes the TI-86 to adjust the values of xMin and xMax and redraw the graph.

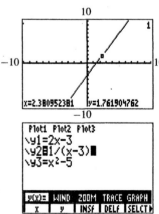

To graph the function

$$y = \frac{1}{x - 3},$$

enter that formula into the "y equals" screen (note the use of parentheses). As before, this example uses the standard viewing window.

For this function, the TI-86 produces the graph shown on the right. This illustrates one of the pitfalls of the connect-the-dots method used by the calculator: The nearly-vertical line segment drawn at $x = 3$ *should not be there*, but it is drawn because the calculator connects the points

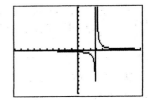

$$x = 2.85714, y = -6.99999 \text{ and } x = 3.01587, y = 62.99999.$$

Calculator users must learn to recognize these flaws in calculator-produced graphs.

The graph of a circle centered at the origin with radius 8 (shown on the square window ZOOM:ZSTD - ZOOM:ZSQR) shows another problem that arises from connecting the dots. When $x = -8.093841$, y is undefined, so no point is plotted (that is, there is no point on this circle that has x-coordinate less than -8, or greater than 8). The next point plotted on the upper half of the circle is $x = -7.824046$ and $y = 1.668619$; since no point had been plotted for the previous x-coordinate, this is not connected to anything, so there appears to be a gap between the circle and the x-axis. The calculator is not "smart" enough to know that the graph should extend from -8 to 8.

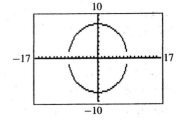

One additional feature of graphing with the TI-86 is that each function can have a "style" assigned to its graph. The symbol to the left of y1, y2, etc. indicates this style, which can be changed by choosing GRAPH:y(x)=:STYLE to cycle through the options. These options are shown on the right (with brief descriptive names); complete details are in the TI-86 manual.

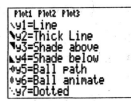

13   Adding programs to the TI-86

The TI-86's capabilities can be extended by downloading or entering programs into the calculator's memory. Instructions for writing a program are beyond the scope of this manual, but programs written by others and downloaded from the Internet (or obtained as printouts) can be transferred to the calculator in one of three ways:

1. If one TI-86 already has a program, it can be transferred to another using the calculator-to-calculator link cable. To do this, first make sure the cable is firmly inserted in both calculators. On the sending calculator, press [2nd][x-VAR] (LINK), then [F1]:[F2] (SEND:PRGM), and then select (by using the [▲] and [▼] keys and [F2]) the program(s) to be transferred. *Before* pressing [F1] (SEND) on the sending calculator, prepare the receiving calculator by pressing [2nd][x-VAR][F2], and *then* press [F1] on the sending calculator.

2. If a computer with the TI-Graph Link is available, and the program file is on that computer (e.g., after having been downloaded from the Internet), the program can be transferred to the calculator using the TI Graph Link software. This transfer is done in a manner similar to the calculator-to-calculator transfer described above; specific instructions can be found in the documentation that accompanies the Graph Link software. (They are not given here because of slight differences between platforms and software versions.)

3. View a listing of the program and type it in manually. (**Note:** Even if the TI-Graph Link cable is not available, the Graph Link software can be used to view program listings on a computer.) While this is the most tedious method, studying programs written by others can be a good way to learn programming. To enter a program, start by choosing [PRGM][F2] (EDIT), then type a name for the new

program (up to eight letters, like "QuadForm" or "Midpoint")—note that the TI-86 is automatically put into ALPHA mode. Then type each command in the program, and press [2nd][EXIT] (QUIT) to return to the home screen when finished.

To run the program, make sure there is nothing on the current line of the home screen, then press [PRGM][F1], select the program using one of the keys [F1]–[F5] and [MORE] (a sample screen is shown; only the first four to six characters of each program name are shown), and press [ENTER]. If the program was entered manually (option 3 above), errors may be reported; in that case, choose GOTO, correct the mistake and try again.

Programs can be found at many places on the Internet, including:

- http://www.awl.com/lhs—the Web site for the text;

- http://www.bluffton.edu/~nesterd—the Web site of the author of this manual;

- http://tifaq.calc.org—A "Frequently Asked Questions" page maintained by Ray Kremer; and

- http://www.ticalc.org.

Examples

Here are the details for using the TI-86 for several of the examples from the textbook. Also given are the keystrokes necessary to produce some of the commands shown in the text's examples. In some cases, some suggestions are made for using the calculator more efficiently.

We first consider examples from the text's Appendices, as the calculator techniques they illustrate are useful throughout the text.

Throughout this section, it is assumed that the textbook is available for reference. The problems from the text are not restated here, and there are frequent references to the calculator screens shown in the text.

Appendix A   Example 1   (page 434)   Solving a Linear Equation

Here is a general discussion of how to use the TI-86 to solve (or confirm solutions for) nearly any equation. We will show multiple approaches for solving the equation $\frac{1}{2}x - 6 = \frac{3}{4}x - 9$. (The answer is $x = 12$.) These procedures can be adapted for any equation, including the one from this example, or those found throughout the text.

There are two graphical methods that can be used to confirm this solution. The first is the **intersection** method. To begin, set up the TI-86 to graph the left side of the equation as y1, and the right side as y2. **Note:** Putting the fractions in parentheses ensures no mistakes with order of operations. This is not crucial for the TI-86, but is a good practice because some other models give priority to implied multiplication. See section 8 of the introduction, page 56.

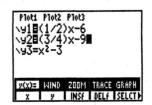

We are looking for an x value that will make the left and right sides of this equation equal to each other, which corresponds to the x-coordinate of the point of intersection of these two graphs.

Next, select a viewing window which shows the point of intersection; we use $[-15, 15] \times [-10, 10]$ for this example. The TI-86 can automatically locate this point using GRAPH:MATH:ISECT (GRAPH MORE F1 MORE F3). Use ▲, ▼ and ENTER to specify which two functions to use (in this case, the only two being displayed), and then use ◄ or ► to specify a guess. After pressing ENTER, the TI-86 will try to find an intersection of the two graphs. The screens below illustrate these steps.

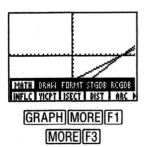   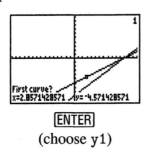   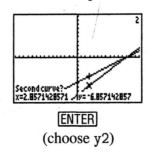   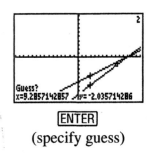

GRAPH MORE F1   ENTER   ENTER   ENTER
MORE F3   (choose y1)   (choose y2)   (specify guess)

The final result of this process is the screen shown on the right. The x-coordinate of this point of intersection is calculated to 14 digits of accuracy, so if the solution were some less "convenient" number (say, $\sqrt{3}$ or $1/\pi$), we would have an answer that would be accurate enough for nearly any computation.

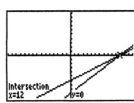

Note: An approximation for the point of intersection can be found simply by moving the TRACE cursor as near the intersection as possible. The amount of error can be minimized by "zooming in" on the graph. This is the only method available for graphing calculators such as the TI-81.

The second graphical approach is to use the x-**intercept method**, which seeks the x-coordinate of the point where a graph crosses the x-axis. Specifically, we want to know where the graph of y1−y2 crosses the x-axis, where y1 and y2 are as defined above. This is because the equation $\frac{1}{2}x - 6 = \frac{3}{4}x - 9$ can only be true when $\frac{1}{2}x - 6 - \left(\frac{3}{4}x - 9\right) = 0$.

To find this x-intercept, begin by defining y3=y1−y2 on the GRAPH:y(x)= screen. We could do this by re-typing the formulas entered for y1 and y2, but having typed those formulas once, it is more efficient to do this as shown on the right. The simplest way to type "y1" and "y2" is to use the [F2] key to produce "y." Note that y1 and y2 have been "de-selected" so that they will not be graphed (see section 12 of the introduction, page 60).

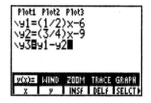

We must first select a viewing window which shows the x-intercept; we again use $[-15, 15] \times [-10, 10]$. The TI-86 can automatically locate this point with the GRAPH:MATH:ROOT ([GRAPH][MORE][F1][F1]) feature ("root" is a synonym for "x-intercept"). The TI-86 prompts for left and right bounds (numbers that are, respectively, less than and greater than the root) and a guess, then attempts to locate the root between the given bounds. (Provided there is only one root between the bounds, and the function is "well-behaved"— meaning it has some nice properties like continuity—the calculator will find it.) The screens below illustrate these steps.

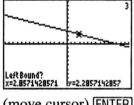 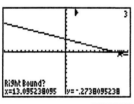 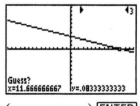 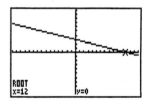

(move cursor) [ENTER]   (move cursor) [ENTER]   (move cursor) [ENTER]   Here is the result.

The TI-86 also offers some non-graphical approaches to solving this equation (or confirming a solution): As illustrated on the right, the TI-86's Solver function attempts to find a value of x that makes the given expression equal to 0, given a guess (10, in this case). The solution is stored in the variable x, but as the screen

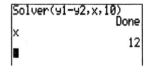

shows, this solution is not automatically displayed. The entry shown use of the fact that y1 and y2 have been defined as the left and right sides of this equation; if that had not been the case, the same results could have been attained by entering (e.g.) Solver((1/2)x-6-((3/4)x-9),x,10). Full details on how to use this function (found in the CATALOG) can be found in the TI-86 manual.

Finally, the TI-86 includes an "interactive solver," accessed with [2nd][GRAPH]. This prompts for the equation to be solved (use [ALPHA][STO▸] to type the equals sign), then allows the user to enter a guess for the solution (or a range or numbers between which a solution should be sought). To solve the equation, place the cursor on the line beginning with x= and press F5.

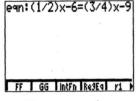

Enter equation

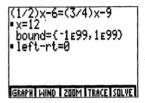

Specify guess, or press F5

Here is the solution.

The solver can also be used with equations containing more than one variable; simply provide values for all but one variable, then place the cursor on the line containing the variable for which a value is needed and press F2.

Note: In this example, we learned how the TI-86 can be used to support an analytic solution. But the TI-86 and any other graphing calculator also can be used for solving problems when an analytic solution is **not** possible—that is, when one cannot solve an equation "algebraically." This is often the case in many "real-life" applications, and is one of the best arguments for the use of graphing calculators.

Appendix A Example 4 (page 436)	Using the Zero-Factor Property
Appendix A Example 5 (page 437)	Using the Square-Root Property
Appendix A Example 6 (page 437)	Using the Quadratic Formula

The TI-86 can solve quadratic equations (as well as higher-degree polynomial equations) using its built-in polynomial solver, accessed through 2nd PRGM (POLY). This first prompts the user for the "order" (degree) of the polynomial, meaning the highest power of x. For the quadratic equations, this should be 2.

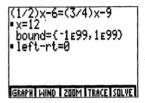

Pressing ENTER then brings up the screen on the right, requesting the coefficients of the equation. Note that the top line of the screen contains a reminder that the expression must be equal to 0. The menu at the bottom indicates that F1 will clear the coefficients, while F5 solves the equation. The entries shown here are for Example 6.

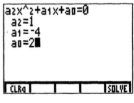

Pressing F5 reports the solutions. F1 allows the user to change the coefficient values (that is, it goes back to the previous screen), and F2 provides a way to store the coefficients in a variable (as a list).

Appendix A   Example 7   (page 438)   Solving a Linear Inequality

Shown is one way to visualize the solution to inequalities like these. The inequality symbols $>$, $<$, $\geq$, $\leq$ are found in the TEST menu ([2nd][2nd][×]). The TI-86 responds with 1 when a statement is true and 0 for false statements. This is why the graph of y1 appears as it does: For values of x less than 4, the inequality is true (and so y1 equals 1), and for $x \geq 4$, the inequality is false (and y1 equals 0). Note that this picture does *not* help one determine what happens when $x = 4$.

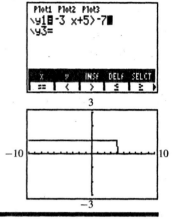

Appendix A   Example 8   (page 439)   Solving a Three-Part Inequality

To see a "picture" of this inequality like the one shown for the previous example requires a little additional work. Setting y1=-2<5+3x≤20 will *not* work correctly. Instead, one must enter either

    y1=(-2<5+3x)(5+3x≤20)   or   y1=(-2<5+3x) and (5+3x≤20)

which will produce the desired results: A function that equals 1 between $-\frac{7}{3}$ and 5, is equals 0 elsewhere. ("and" is most easily entered either by selecting it from the CATALOG, or by simply typing it letter-by-letter; note the spaces before and after, typed by pressing [2nd][(-)]).

Appendix B   Example 2   (page 444)   Using the Midpoint Formula

The TI-86 can do midpoint computations nicely by putting coordinates in parentheses, as shown on the right. (The TI-86 interprets an ordered pair such as (8,-4) as the complex number $8 - 4i$, but since adding two complex numbers means adding their corresponding parts, the computations are done in the correct way to find the midpoint.)

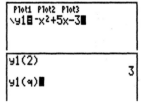

Appendix C   Example 5   (page 455)   Using Function Notation

The TI-86 will do *some* computations in function notation. Simply enter the function in y1 (or any other function). Return to the home screen and enter y1(2) to evaluate that function with the input 2. The result shown on the screen agrees with (a) in the text.

The TI-86 is less helpful if asked to evaluate expressions like those given in Example 5(b). For example, "y1(q)" produces an error message or an unpredictable result (depending on how the variable q is defined), rather than the desired result $-q^2 + 5q - 3$.

The table features of the TI-86 allow another method of computing function values. To use these features, begin by entering the formula on the GRAPH:y(x)= screen, as one would to create a graph. (The highlighted equals signs determine which formulas will be displayed in the table, just as they do for graphs.)

Next, press [TABLE][F2] to access the TABLE SETUP screen. The table will display *y*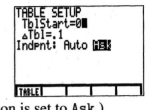
values for given values of *x*. For this example, we want to set Indpnt to Ask,
which means that the TI-86 will wait for us to enter a value for *x* (the "independent"
variable), then display the corresponding function value(s). When Indpnt is set
to Auto, the TI-86 automatically generates the values of *x*, starting at TblStart
and with a step size of ΔTbl. (TblStart and ΔTbl are ignored if the Indpnt option is set to Ask.)

When the TABLE SETUP options are set satisfactorily, press [F1] (TABLE) to produce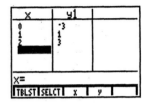
the table. In the screen shown, values of y1 are shown for the three input values
0, 1, and 2. (These three input values had to be individually entered.)

To get graphical confirmation, enter the appropriate formula for y1 and graph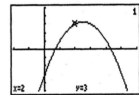
in any window that includes *x* = 2. Press TRACE ([GRAPH][F4]), then press the
[▶] and [◀] keys to change the value of *x*. It may not be possible to make *x*
exactly equal to 2 in this manner (see section 12 of the introduction, page 60),
but rough confirmation that *f*(2) = 3 can be found by observing that *y* is close
to 3 when *x* is close to 2.

The TI-86 makes it possible to trace to any real number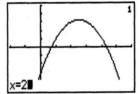
value for *x* between xMin and xMax. Simply type a number
or expression (like 1/π or √2) while in TRACE mode. The
number appears at the bottom of the window in a larger font
size than the TRACE coordinates. Pressing [ENTER] causes
the TRACE cursor to jump to that *x*-coordinate. This same result can be achieved using the GRAPH:EVAL
command ([GRAPH][MORE][MORE][F1]).

Section 1.1	Example 2	(page 4)	Calculating with Degrees, Minutes, and Seconds

Section 1.1	Example 3	(page 4)	Converting between Decimal Degrees and Degrees, Minutes, and Seconds

See section 9 of this chapter's introduction (page 56) for information about selecting Degree mode. The
alternative to putting the calculator in Degree mode is to use the degree symbol following each angle measure.

The MATH:ANGLE menu ([2nd][×][F3]) is shown on the right. It includes the degrees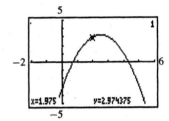
and minutes symbols, and the ▶DMS operator (which causes an angle to be displayed
in degrees, minutes, and seconds, rather than as a decimal).

Entering angles on a TI-86, however, is different than on a TI-83 (which was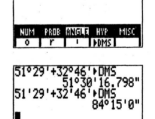
used to create the screens shown in the text.) The degree and seconds symbols
(° and ") are not used; the minutes symbol is used for all three positions, as in
the screen on the right. (The results of the computations are displayed with the
"usual" symbols.)

The screen on the right shows the proper way to enter the angles from Example 3 for computation on the TI-86.

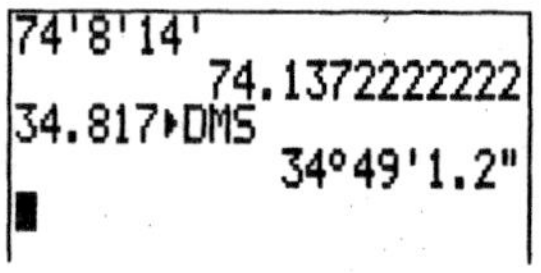

```
74'8'14'
          74.1372222222
34.817▶DMS
            34°49'1.2"
■
```

Section 1.2   Example 1   (page 10)   Finding Angle Measures

See page 63 for general information about solving equations using the TI-86. (Of course, one must use knowledge from geometry to obtain the equation in the first place.)

Section 1.3   Example 4   (page 23)   Finding Function Values of Quadrantal Angles

The alternative to putting the calculator in Degree mode is to use the degree symbol ([2nd][×][F3][F1]) following each angle measure; e.g., enter sin 90° rather than just sin 90. (However, see the comment in the next example.)

Since the cotangent, secant, and cosecant functions are the reciprocals of the tangent, cosine, and sine, they can be entered as (e.g.) 1/sin 90. (This is illustrated in Figure 31 on page 27 of the text.) Note, though, that this will not properly compute cot 90°, since 1/tan 90 produces a domain error. Entering cot x as cos x/ sin x will produce the correct result at 90°.

One might guess that the other three trigonometric functions are accessed with [2nd] followed by [SIN], [COS], or [TAN] (which produce, e.g., sin^{-1}). This is **not** what these functions do; in this case, the exponent -1 does not mean "reciprocal," but instead indicates that these are inverse functions (which are discussed in Sections 2.3 and 6.1 of the text). The text comments on this distinction at the bottom of page 27.

Section 2.3   Example 1   (page 62)   Finding Function Values with a Calculator

Recall that the TI-86 uses a slightly different format for entering angles in degrees and minutes, as illustrated on the right. This screen also shows how computations like the secant in part (b) can be entered on a single line. These computations were done in Degree mode.

```
sin 49'12'
          .756995055652
(cos 97.977)⁻¹
          -7.2058792129
■
```

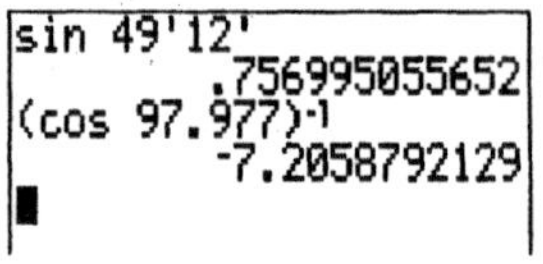

The screen on the right (with computations done in Radian mode) illustrates a somewhat unexpected behavior: Even if an angle is entered in DMS format, the TI-86 assumes that the angle is in radians. In order to remedy this, either put the calculator in Degree mode, or use the degree symbol ([2nd][×][F3][F1]) as was done in the second entry.

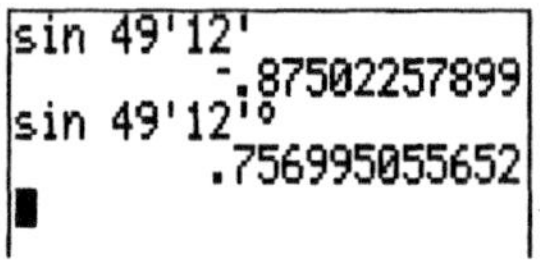

```
sin 49'12'
         -.87502257899
sin 49'12'°
          .756995055652
■
```

Section 2.3   Example 2   (page 62)   Using an Inverse Trigonometric Function to Find an Angle

The sin^{-1} ("inverse sine," or "arcsine") function is [2nd][SIN], while cos^{-1} is [2nd] [COS]. The computations shown in Figure 16 in the text were done in Degree mode; the screen on the right shows the result for Example 7 when done in Radian mode. Note that in the second entry, an attempt was made to get the TI-86 to report the result in degrees (by placing the degree symbol at the end of the entry), but this does not have the desired result.

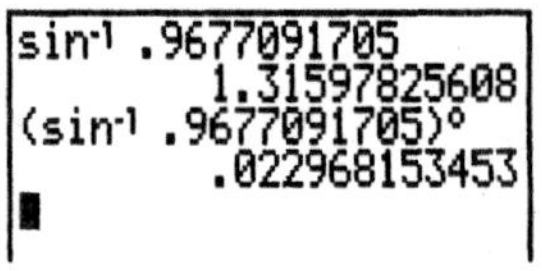

```
sin⁻¹ .9677091705
          1.31597825608
(sin⁻¹ .9677091705)°
          .022968153453
■
```

Section 2.4   Example 1   (page 69)   Solving a Right Triangle Given an Angle and a Side

---

Section 2.4   Example 2   (page 70)   Solving a Right Triangle Given Two Sides

For problems like these, the Ans variable can be used to avoid loss of accuracy from rounding off intermediate results. Shown are computations for Example 1. (Be sure the TI-86 is in Degree mode. Also, recall that the arcsine (or inverse sine) function is [2nd][SIN] — *not* [SIN][2nd][EE].)

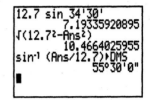

Section 2.5   Example 4   (page 79)   Solving a Problem Involving Angles of Elevation

The TI-86's GRAPH:MATH:ISECT ([GRAPH][MORE][F1][MORE][F3]) feature will automatically locate the intersection of two graphs. This feature was previously illustrated on page 63, but we repeat the description here: Use [▲], [▼] and [ENTER] to specify which two functions to use (in this case, the only two being displayed), and then use [◄] or [►] to specify a guess. After pressing [ENTER], the TI-86 will try to find an intersection of the two graphs. The screens below illustrate these steps; the final result is essentially the same as the screen shown in text Figure 32.

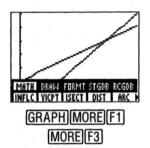
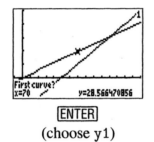
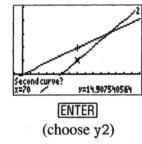
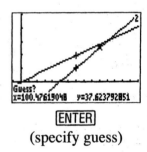

| [GRAPH][MORE][F1] [MORE][F3] | [ENTER] (choose y1) | [ENTER] (choose y2) | [ENTER] (specify guess) |

Section 3.1   Example 1   (page 95)   Converting Degrees to Radians

The number π is available as [2nd][^], and the degree symbol is [2nd][×][F3][F1]. With the calculator in Radian mode (see page 56), entering 45° causes the TI-86 to automatically convert to radians.

A useful technique to aid in recognizing when an angle is a multiple of π is to divide the result by π. This approach is illustrated in the screen on the right, showing that 45° is $\pi/4$ radians, and 30° is $\pi/6$ radians. This screen also makes use of the ►Frac command from the MATH:MISC menu ([2nd][×][F5][MORE][F1]), which simply means "display the result of this computation as a fraction, if possible." This is a useful enough command that one may wish to put it in the CUSTOM menu (see page 56).

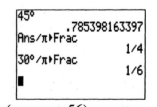

An alternative to using the degree symbol is to store $\pi/180$ in the calculator variable D (see page 56). Then typing, for example, 45D [ENTER] will multiply 45 by $\pi/180$. This approach will work regardless of whether the calculator is in Degree or Radian mode. (A value stored in a variable will remain there until it is replaced by a new value.)

Section 3.1   Example 2   (page 95)   Converting Radians to Degrees

With the TI-86 in Degree mode (see page 56), the radian symbol (a superscripted r), produced with 2nd ×
F3 F2, will automatically change a radian angle measurement to degrees.

Alternatively, with the value $180/\pi$ stored in the calculator variable R (see
page 56), typing (9π/4)R ENTER will convert from radians to degrees regardless of whether the calculator is in Degree or Radian mode. (The same result can
be achieved by *dividing by* the calculator variable D as defined in the previous
example.)

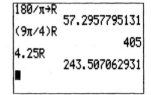

Section 3.3   Example 4   (page 112)   Finding a Number Given Its Circular Function Value

The arccosine (or inverse cosine) function is 2nd COS —*not* COS 2nd EE. Likewise, $\tan^{-1}$ is 2nd TAN.

Section 4.1   Example 1   (page 135)   Graphing $y = a \sin x$

Note that the TI-86 must be in Radian mode in order to produce the desired graph. See page 61 for
information on setting the thickness of a graph.

The graphs in the text are shown in the "trig viewing window," described on the top of page 135. The
TI-86's default trig window (selected with GRAPH F3 MORE F3) is slightly wider than that shown in the
text; it shows x from -2.625π to 2.625π. (If the TI-86 is in Degree mode, xMin and xMax will be ±472.5
instead of $\pm2.625\pi$.) The graph below is shown using the text's trig window, rather than the TI-86's.

It is possible to distinguish between the two graphs without having them drawn
using different styles by using the TRACE (GRAPH F4) feature. On the right, the
trace cursor is on graph 2—that is, the graph of $y_2 = \sin x$.

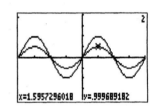

Section 4.2   Example 6   (page 151)   Modeling Temperature with a Sine Function

Given a set of data pairs (x, y), the TI-86 can produce a scatter diagram (like the points shown in Figure 19)
and can find various formulas (including linear and quadratic, as well as more complex formulas like a
sine function) that approximate the relationship between x and y. These formulas are called "regression
formulas."

The first step is to enter the data into the TI-86. This is done by pressing 2nd +
F2 (STAT:EDIT), which should produce the screen on the right. If xStat, yStat,
and fStat are not on the STAT:EDIT screen, and pressing ◄ and ► does not reveal
them, they can be restored to this screen by entering the command SetLEdit ("set
up the list editor") on the home screen. This command is available by pressing
2nd – F5 MORE MORE MORE F3.

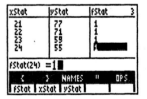

For the scatter diagram shown in the text, enter the temperatures twice—24 pairs of number altogether. In
the STAT:EDIT screen, enter the month numbers (1–24) into the first column (xStat) and the temperatures
into the second column (yStat). The third column (fStat) should contain 24 "1"s; this tells the TI-86

that each of these (x, y) pairs occurs only once in the data list. If any column already contains data, the ⌊DEL⌋ key can be used to delete numbers one at a time, or—to delete the whole column at once—press the ⌊▲⌋ key until the cursor is at the top of the column (on xStat, yStat, or fStat) and press ⌊CLEAR⌋⌊ENTER⌋. Make sure that all three columns contain the same number of entries. The screen above shows the final entry for the data of this example.

To produce the scatter diagram, press ⌊2nd⌋⌊+⌋⌊F3⌋ to bring up the STAT:PLOT menu, shown on the right. Select Plot1 by pressing ⌊F1⌋ (or choose one of the other two plots).

Make the settings shown on the screen on the right. For Type, choose ⌊F1⌋ (SCAT) for a scatter diagram.

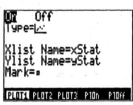

Next, check that nothing else will be plotted: Press ⌊GRAPH⌋⌊F1⌋ and make sure that only Plot1 is highlighted. If Plot2 or Plot3 (or an equals sign) is highlighted, use the arrow keys to move the cursor to it, then press ⌊F5⌋ (SELCT).

Finally, set up the viewing window (GRAPH:WINDOW) as shown in Figure 19 of the text—or press ⌊GRAPH⌋ ⌊F3⌋⌊MORE⌋⌊F5⌋ (GRAPH:ZOOM:ZDATA), which automatically adjusts the window to show all the data in the plot. This should produce a plot like that shown in the text.

Note: When finished with a statistics plot like this one, it is a good idea to turn it off so that the TI-86 will not attempt to display it the next time ⌊GRAPH⌋ is pushed. This can be done using the SELCT option on the GRAPH:y(x)= screen, or by executing the PlOff command, by pressing ⌊2nd⌋⌊+⌋⌊F3⌋⌊F5⌋⌊ENTER⌋.

To find the regression equation, press ⌊2nd⌋⌊+⌋⌊F1⌋ (STAT:CALC). The bottom of the screen now lists the various options for the type of calculation to be done. Pressing ⌊MORE⌋⌊F2⌋ tells the calculator to perform a "SinR"—a linear regression. (The command SinR is placed on the home screen.

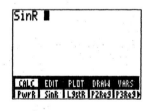

Pressing ⌊ENTER⌋ displays the results of the SinR (shown on the right; the TI-86 had been set to display two digits after the decimal). These numbers nearly agree with those shown in the text (Figure 21); the TI-86 uses a slightly different method to find this formula,so its results are not identical to those given by the TI-83.

Section 4.3   Example 1   (page 158)   Graphing $y = a\sec bx$

Section 4.3   Example 2   (page 159)   Graphing $y = a\csc(x - d)$

See section 9 of the introduction (page 56) for information about DrawLine versus Dot (DrawDot) mode. The function in Example 1 can be entered as y1=2cos (x/2)⁻¹ (or as shown in the text). The function in Example 2 can be entered as y1=(3/2)(1/sin (x-π/2)) or y1=3/(2sin (x-π/2)). A reminder: sin⁻¹ (2nd SIN) is *not* the cosecant function.

Section 5.1   Example 3   (page 185)   Rewriting an Expression in Terms of Sine and Cosine

The top screen on the right shows how these expressions are entered on the "y equals" screen. As an alternative to graphing these two functions, the TI-86's table feature (see page 66 of this manual) can be used: If the y values are the same for a reasonably large sample of x values, one can be fairly sure (though not certain) that the two expressions are equal. To make this approach more reliable, be sure to choose x values that are not, for example, all multiples of π.

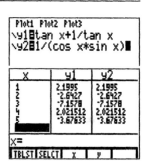

Section 5.3   Example 1   (page 198)   Finding Exact Cosine Function Values

---

Section 5.4   Example 1   (page 206)   Finding Exact Sine and Tangent Function Values

The TI-86 can graphically and numerically support exact value computations such as $\cos 15° = \frac{\sqrt{6}+\sqrt{2}}{4}$. Starting with a graph of y1=cos x, the TI-86 makes it possible to evaluate y1 (or any function) for any value of x between xMin and xMax using the TI-86's GRAPH:EVAL (GRAPH

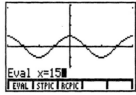
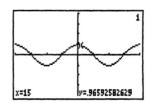

MORE MORE F1) option. GRAPH:EVAL prompts for a value of x, which can be either a number like 15 or expression (like $1/\pi$ or $\sqrt{2}$). Pressing ENTER causes the cursor to jump to that x-coordinate.

Alternatively, a table of values like those shown here can be used to find the value of cos 15°. Of course, the screen shown in the text in support of Example 1(b) shows the other part of this process: Computing the decimal value of $(\sqrt{6}+\sqrt{2})/4$ and observing that it agrees with those found here.

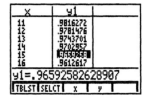

Note that Example 1 in Section 5.6 (page 222) shows that cos 15° can also be written as $\frac{\sqrt{2+\sqrt{3}}}{2}$.

| Section 6.1 | Example 1 | (page 238) | Finding Inverse Sine Values |

| Section 6.1 | Example 2 | (page 239) | Finding Inverse Cosine Values |

Of course, it is not necessary to graph $y = \sin^{-1} x$ to find these values; one can simply enter, e.g., sin⁻¹ (1/2) on the home screen. The first entry of the screen on the right shows what happens when the calculator is in Degree mode; note that the result is not in $[-\pi/2, \pi/2]$. With the calculator in Radian mode, results similar to those in the text are found, and the method employed on page 69 of this manual (in the discussion of text Example 1 from Section 3.1) confirms that these values are $\pi/6$ and $3\pi/4$.

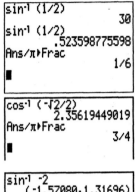

Note that the TI-86 does *not* give an error for the input sin⁻¹ -2, but instead gives a complex result: $-\pi/2 + i \ln(2 + \sqrt{3})$. The reason that this is technically a correct result is beyond the scope of the textbook; TI-86 owners should recognize that a complex answer is not appropriate for this problem, and so should ignore this result.

| Section 6.1 | Example 4 | (page 243) | Finding Inverse Function Values with a Calculator |

Note that the answer given for (c), 109.499054°, overrepresents the accuracy of that value. A typical rule for doing computations involving decimal values (like −0.3541) is to report only as many digits in the result as were present in the original number—in this case, four. This means the reported answer should be "about 109.5°," and in fact, any angle θ between about 109.496° and 109.501° has a cotangent which rounds to −0.3541. (See the discussion of significant digits on page 68 of the text.)

| Section 6.2 | Example 6 | (page 253) | Describing a Musical Tone from a Graph |

| Section 6.3 | Example 5 | (page 259) | Analyzing Pressures of Upper Harmonics |

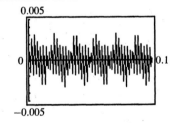

Note that the calculator screens shown in Figures 23–26 illustrate the importance of choosing a "good" viewing window. If we choose the wrong vertical scale (yMin and yMax), we might not be able to see the graph at all—it might be squashed against the *x*-axis. If we make the window too wide—that is, if xMax minus xMin is too large—we might see the "wrong" picture, like the one on the right (for Example 5 from Section 6.3): We seem to see six periods in this view, when in fact there are 44. Each of the six "pseudo-cycles" is made of parts of seven or eight full periods.

This observation—that a periodic function, viewed at fixed intervals, can appear to be a *different* periodic function—is the same effect that causes wagon wheels to appear to run backwards in old movies.

Section 7.4   Example 1   (page 307)   Finding Magnitude and Direction Angle

Section 7.4   Example 2   (page 307)   Finding Horizontal and Vertical Components

The TI-86 does not have the conversion functions shown in Figures 26 and 28, but can do the desired conversions in a manner that is perhaps even more convenient. The TI-86 recognizes vectors entered in either of two formats:

[*horizontal component*, *vertical component*]   or   [*magnitude ∠ angle*]

(The square brackets are 2nd[(] and 2nd[)], and "∠" is 2nd[,].) Regardless of how the vector is entered, the TI-86 displays it according to the RectV CylV SphereV mode setting (see page 56); specifically, it displays the vector in component form in RectV mode, and in magnitude/angle form for either of the other two modes.

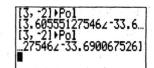

For Example 1, then, either put the TI-86 in CylV or SphereV mode, or (perhaps more conveniently), use the ▶Pol command (2nd[8][F4][F3]), as the screen on the right illustrates. This causes a vector to be displayed in magnitude/direction angle format regardless of the mode setting. On the fourth line, we see the results of pressing [▶] to see all the digits of the angle. (With the TI-86 in Degree mode, the angle returned is in degrees.)

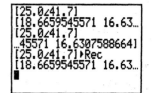

For Example 2, both the horizontal and vertical components can be found at once, as the screen on the right shows. (The vertical component is too long to fit on the screen, but the other digits can be seen by pressing [▶]. Note that the screen shown in the text was created with the calculator set to display only one digit after the decimal.) These computations were done with the TI-86 in Degree mode, so it was not necessary to include the degree symbol on the angle. The third entry shown uses the ▶Rec command, which forces the vector to be displayed in component form regardless of the mode setting. (It was unnecessary in this case, because the TI-86 with in RectV mode.)

Finally, observe that this conversion could be done by typing 25cos(41.7) and 25sin(41.7); which approach to use is a matter of personal preference.

Section 7.4   Example 3   (page 308)   Writing Vectors in the Form (*a*, *b*)

Note how easily these conversions are done in the TI-86's polar vector format. (The TI-86 was in Degree mode.)

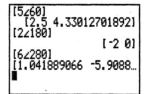

Section 7.4   Example 5   (page 309) Performing Vector Operations

---

Section 7.4   Example 6   (page 310) Finding Dot Products

Using the TI-86's vector notation, the vectors can be stored in calculator variables (see page 56) which can then be used to do the desired operations. Shown are the commands to compute 5(c): $4\mathbf{u} - 3\mathbf{v}$, and 6(a): $\langle 2, 3 \rangle \cdot \langle 4, -1 \rangle$ (the dot command is 2nd 8 F3 F4).

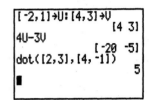

Section 8.1   Example 1   (page 333) Writing $\sqrt{-a}$ as $i\sqrt{a}$

Section 8.1   Example 4   (page 334) Finding Products and Quotients Involving Negative Radicands

The TI-86 does these computations almost as shown in the text, except that it reports the results differently. The screen on the right illustrates the output from the computations for parts (b) and (d) of Example 4. The TI-86 uses the format (a, b) for the complex number $a + bi$; whenever the TI-86 does a computation involving square roots of negative numbers, it reports the results in this format *even if the result is real*. Thus $(-7.74596669242, 0)$ represents the real number $-7.74596669242 \approx -2\sqrt{15}$, and $(0, 1.41421356237)$ is $i\sqrt{2}$.

The calculator always reports complex *results* in this format, or in polar format, which is similar to the text's trigonometric format. However, it can recognize *input* typed in using "$a + bi$" format by first defining a variable i as the complex number $(0, 1) = 0 + 1i$. Shown on the right is a computation using this definition. One could also use I (uppercase), which can be typed with one less keystroke, but i will be used in these examples.

Section 8.1   Example 2   (page 333) Solving Quadratic Equations for Complex Solutions

---

Section 8.1   Example 3   (page 334) Solving a Quadratic Equation for Complex Solutions

The TI-86's POLY feature can find complex solutions as well as real ones. In the screens on the right, the TI-86 was set to display four digits after the decimal.

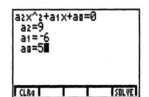 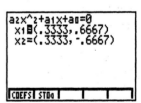

| Section 8.1 | Example 5 | (page 335) | Adding and Subtracting Complex Numbers |

| Section 8.1 | Example 6 | (page 336) | Multiplying Complex Numbers |

| Section 8.1 | Example 7 | (page 337) | Simplifying Powers of *i* |

| Section 8.1 | Example 8 | (page 338) | Dividing Complex Numbers |

Using the variable i defined to be (0, 1) — as was described in the previous example — computations like these can be done in a manner similar to those illustrated in the text's calculator screens. The screens on the right show (in the TI-86's format) the same results as those in the text accompanying Examples 6 and 8.

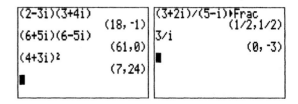

| Section 8.2 | Example 2 | (page 342) | Converting from Trigonometric Form to Rectangular Form |

Complex numbers can be entered in trigonometric form in a manner similar to the polar format for vectors (see page 74); the screen on the right illustrates how to quickly convert from the TI-86's trigonometric form to its rectangular form (assuming the calculator is in RectC mode; see page 56). "∠" is [2nd][,].

| Section 8.2 | Example 3 | (page 343) | Converting from Rectangular Form to Trigonometric Form |

Although the TI-86 does not have the conversion functions shown in the text, the same thing can be accomplished with the abs and angle commands, found in the TI-86's CPLX menu ([2nd][9]), and shown on page 347 of the text. On the screen on the right, note that if one makes use of the variable i defined as (0, 1), the angle and abs functions require parentheses; the third output is incorrect because it first computes the "angle" of the real number $-\sqrt{3}$ (which is π) and then adds this to i.

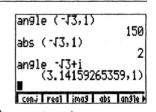

More convenient is the TI-86's ▸Pol command, found by pressing [MORE] in the CPLX menu. (This command is also found in the VECTR:OPS menu.) This gives both the magnitude and angle at the same time.

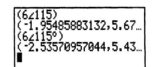

| Section 8.2 | Example 4 | (page 344) | Converting Between Trigonometric and Rectangular Forms Using Calculator Approximations |

Aside from entering the expression in (a) as it is shown in the text (using the cos and sin functions), one can use the TI-86's polar format $(r \angle \theta)$, but be sure to either put the TI-86 in the correct mode (Degree or Radian) or use the degrees or radians symbol from the MATH:ANGLE menu. Both computations on the right were done in Radian mode, and the first result is incorrect.

Section 8.3   Example 1   (page 348)   Using the Product Theorem

Section 8.3   Example 2   (page 349)   Using the Quotient Theorem

Section 8.3   Example 3   (page 349)   Using the Product and Quotient Theorems with a Calculator

The TI-86's polar complex format makes these computations very easy to enter.

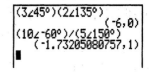

Section 8.4   Example 1   (page 353)   Finding a Power of a Complex Number

The screen on the right shows several options for computing $(1 + i\sqrt{3})^8$ with the TI-86. The first is fairly straightforward, but the reported result shows the complex part of the answer ($128i\sqrt{3}$) given in decimal form. For an "exact" answer, the TI-86's polar format can be used. The TI-86 was in Degree mode for the first of the two polar-format answers, and in Radian mode for the second (the angle given in the second answer is $2\pi/3$).

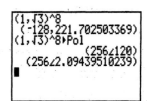

Section 8.4   Example 4   (page 356)   Solving an Equation by Finding Complex Roots

The TI-86's POLY polynomial solver (see page 65) can find complex solutions as well as real ones, and can be used to see the pattern of these roots, if the calculator is in PolarC and Degree modes. The two negative angles shown in the second screen are coterminal with the angles $216°$ and $288°$ given in the text.

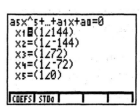

Section 8.5   Example 3   (page 361)   Examining Polar and Rectangular Equations
of Lines and Circles

Section 8.5   Example 4   (page 363)   Graphing a Polar Equation (Cardioid)

Section 8.5   Example 5   (page 363)   Graphing a Polar Equation (Rose)

Section 8.5   Example 6   (page 364)   Graphing a Polar Equation (Lemniscate)

To produce these polar graphs, the TI-86 should be set to Degree and Polar modes (see the screen on the right). In this mode, GRAPH F1 allows entry of polar equations (r as a function of θ). One could also use Radian mode, adjusting the values of θMin, θMax, and θStep accordingly (e.g., use 0, 2π, and $\pi/30$ instead of 0, 360, and 5).

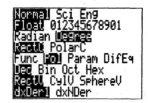

For the cardioid, rose, and lemniscate, the window settings shown in the text show these graphs on "square" windows (see section 11 of the introduction, page 59), so one can see how their proportions compare to those of a circle. Note, however, that on the TI-86, these window are not square; the ZOOM:ZSQR option can be used to adjust the window dimensions to make it square.

For the cardioid, the value of θStep does not need to be 5, although that choice works well for this graph. Too large a choice of θStep produces a graph with lots of sharp "corners," like the one shown on the right (drawn with θStep=30). Setting θStep too small, on the other hand, produces a smooth graph, but it is drawn very slowly. Sometimes it may be necessary to try different values of θStep to choose a good one.

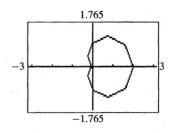

The lemniscate can be drawn by setting θMin=0 and θMax=180, or θMin=-45 and θMax=45. In fact, with θ ranging from -45 to 225, the graph of r1=$\sqrt{\cos (2\theta)}$ (alone) will produce the entire lemniscate. (θStep should be about 5.)

The rose can be produced by setting θMin=0 and θMax=360, or using any $360°$-range of θ values (with θStep about 5).

Section 8.5   Example 7   (page 365)   Graphing a Polar Equation (Spiral of Archimedes)

To produce this graph on the viewing window shown in the text, the TI-86 must be in Radian mode. (In Degree mode, it produces the same shape, but magnified by a factor of $180/\pi$ — meaning that the viewing window needs to be larger by that same factor.)

Section 8.6   Example 1   (page 371)   Graphing a Plane Curve Defined Parametrically

Place the TI-86 in Parametric mode, as the screen on the right shows. In this mode, GRAPH F1 allows entry of pairs of parametric equations (x and y as functions of t). No graph is produced unless both functions in the pair are entered and selected (that is, both equals signs are highlighted).

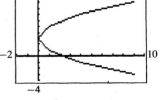

The value of tStep does not need to be 0.05, although that choice works well for this graph. Too large a choice of tStep produces a less-smooth graph, like the one shown on the right (drawn with tStep=1). Setting tStep too small, on the other hand, produces a smooth graph, but it is drawn very slowly. Sometimes it may be necessary to try different values of tStep to choose a good one.

Section 8.6   Example 3   (page 372)   Graphing a Plane Curve Defined Parametrically

This curve can be graphed in Degree mode with tMin=0 and tMax=360, or in Radian mode with tMax=2π. In order to see the proportions of this ellipse, it might be good to graph it on a square window. This can be done most easily with the ZOOM:ZSQR option (GRAPH F3 MORE F2). On a TI-86, initially with the window settings shown in the text, this would result in the window $[-6.8, 6.8] \times [-4, 4]$.

Section 8.6   Example 5   (page 373)   Graphing a Cycloid

The TI-86 *must* be in Radian mode in order to produce this graph.

Section 8.6   Example 6   (page 374)   Simulating Motion with Parametric Equations

Section 8.6   Example 8   (page 375)   Analyzing the Path of a Projectile

Parametric mode is particularly nice for analyzing motion, because one can picture the motion by watching the calculator create the graph, or by using TRACE ([GRAPH][F4]) and watching the motion of the trace cursor. (When tracing in parametric mode, the [▶] and [◀] keys increase and decrease the value of t, and the trace cursor shows the location (x, y) at time t.) Figure 40 illustrates tracing on the projectile path in Example 8. Note that the value of t changes by $\pm$tStep each time [▶] or [◀] is pressed, so obviously the choice of tStep affects which points can be traced.

The TI-86's graph styles (see page 61) can be useful, too. The screen on the right shows the three paths for Example 6 being plotted in "ball path" style.

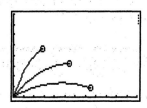

Section 9.1   Example 4   (page 393)   Using a Property of Exponents to Solve an Equation

Below the calculator screen shown in the text, the caption refers to "the x-intercept method of solution." This and other methods for solving equations were described beginning on page 63 of this manual.

Section 9.1   Example 11   (page 397)   Using Data to Model Exponential Growth

The scatter diagram in Figure 10(a) and the "exponential regression" in Figure 12 can be reproduced by adapting the procedures described on page 70 of this manual. The corresponding command on the TI-86 is ExpR ([2nd][+][F1][F5]).

The two calculator screens in Figure 11 use the "intersection method" of solving equations; see page 63 for a description.

Section 9.3   Example 1   (page 414)   Finding pH

For (a), the text shows $-\log(2.5*{10}^{\wedge}(-4))$, but this could also be entered as shown on the first line of the screen on the right, since "E" (produced with [EE]) and "*10^" are nearly equivalent. The two are not completely interchangeable, however; in particular, in part (b), "10^" **cannot** be replaced with "E", because "E" is only valid when followed by an *integer*. That is, E-7 produces the same result as 10^-7, but the last line shown on the screen produces a syntax error.

(Incidentally, "10^" is [2nd][LOG], but [1][0][^] produces the same results.)

Introduction

The information in this section is essentially a summary of material that can be found in the TI-89 manual. Consult that manual for more details.

Owners of a TI-92 will find that most of this material applies to that calculator as well.

1   Power

To power up the calculator, simply press the ON key. The screen displayed at this point depends on how the TI-89 was last used. It may show the "home screen"—a menu (the toolbar) across the top, the results of previous computations (if any) in the middle (the history area), a line for new entries (which may be blank, or may show the previous entry), and a status line at the bottom. An example of this home screen is shown on the right.

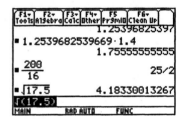

The TI-89 may show some other screen—perhaps a graph, an error message, a menu, or something else. If so, one can return to the home screen by pressing either the HOME key (from a graph), or the ESC key (from an error message), or 2nd ESC (from a menu). Note that the "second function" of ESC—written in yellow type above the key—is "QUIT."

If the screen is blank, or is too dark to read, one may need to adjust the contrast (see the next section).

To turn the calculator off, press 2nd ON (OFF), in which case the TI-89 will start up at the home screen next time. This will not work if an error message is displayed. Pressing ◆ ON will also turn the TI-89 off, but when ON is next pressed, the screen will show exactly what it showed before. The calculator will automatically shut off if no keys are pressed for several minutes, in which case it will behave as if ◆ ON had been pressed.

2   Adjusting screen contrast

If the screen is too dark (all black), decrease the contrast by pressing and holding ◆ and ⊟. If the screen is too light, increase the contrast by pressing and holding ◆ and ⊞. If the screen never becomes dark enough to see, the batteries should be replaced.

3   Replacing batteries

To replace the four AAA batteries, first turn the calculator off (2nd ON), then remove the back cover, remove and replace each battery, replace the back cover, then turn the calculator on again. (After replacing batteries, one may need to adjust the contrast down as described above.) Note: The status line at the bottom of the home screen should display BATT when the batteries are getting low.

4   Basic operations

Simple computations are entered in essentially the same way they would be written. For example, to compute $2 + 17 \times 5$, press 2 + 1 7 × 5 ENTER (the ENTER key tells the calculator to act on what has been typed). Standard order of operations (including parentheses) is followed. Note that the entry and the result (87) are displayed in the last line of the history area, and the entry is also displayed (and highlighted) in the entry line. If any new text is typed, this highlighted text will be deleted.

The result of the most recently entered expression is stored in ans (1), which is typed by pressing 2nd (-) (the word "ANS" appears in yellow above this key). For example, 5 + 2nd (-) ENTER will add 5 to the result of the previous computation. Note that in the history area, "ans(1)" has been replaced by "87."

After pressing ENTER, the TI-89 automatically produces ans (1) if the first key pressed is one which requires a number before it; the most common of these are +, −, ×, ÷, ∧, and STO▸. For example, + 5 ENTER would accomplish the same thing as the keystrokes above (that is, it adds 5 to the previous result). Again, note that the history area shows the value of ans (1) rather than the text "ans (1)."

Although the previously entered expression disappears from the entry line if anything is typed, that expression can be re-evaluated by simply pressing ENTER. This can be especially useful in conjunction with ans (1).

Several expressions can be evaluated together by separating them with colons (2nd 4). When ENTER is pressed, the result of the *last* computation is displayed. (The other results are lost. An example showing how this can be used is shown later.)

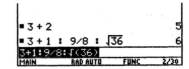

5   Cursors

When typing, the appearance of the cursor and the status line indicates the behavior of the next keypress.

The cursor appears as either a flashing vertical line (the default) or a flashing solid block. The vertical line indicates that subsequent keypresses will be *inserted* at the current cursor location. The block cursor indicates that subsequent keypresses will *overwrite* the character(s) to the right of the cursor. (Of course, if the cursor is located at the right end of the entry line text, these two behaviors are equivalent.) To switch between between these two modes of operation, press [2nd][←] (INS).

By default, most keys produce the character shown on the key itself. The four modifier keys [2nd], [♦], [↑], and [alpha] change this. Pressing any one of these keys causes a corresponding indicator to appear in the status line, and the next keypress will then do something different from its primary function. Pressing [2nd] or [♦] causes the next keypress to produce the results—the character or operation—indicated by (respectively) the yellow or green text above that key. If [2nd] or [♦] is pressed by mistake, pressing it a second time will undo that modifier.

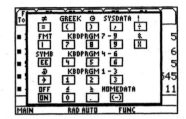

For [2nd], that makes each key's function fairly clear, but many of the keys have no green text above them, leading one to think that the [♦] modifier would accomplish nothing with that key. In fact, nearly every key does something in response to the [♦] key. For an easy way to see the [♦] functions of the lower half of the keypad, press [♦][EE], which produces the display on the right. (Note that the letter associated with [EE] is "K"—think "K" for "keys.") For example, [♦] followed by [=], [)], [÷], [×], [STO▸], [0], or [.] produces the character shown. [♦][(] followed by any letter produces the Greek equivalent of that letter (or as near an equivalent as there is); e.g., [♦][(][Z] produces ζ ("zeta"). [♦][)] allows one to change the number of previous entries saved in the history area. The other [♦] functions are beyond the scope of this manual.

Pressing [alpha] means that the next keypress will produce the (lowercase) letter or other character printed in purple above that key (if any). [alpha] has no effect on (and is not needed for) [X], [Y], [Z], and [T]. Following that letter, subsequent keypresses will produce their primary functions (i.e., not letters). To produce an uppercase letter, press [alpha][↑] followed by a letter key. ([alpha] is not needed for [X], [Y], [Z], and [T].)

The TI-89 can be "locked" into (lowercase) alphabetic mode by pressing [2nd][alpha] (or [alpha][alpha]). From then on, each key produces its letter. This continues until [alpha] is pressed again, which takes the TI-89 out of alphabetic mode.

To lock the TI-89 in uppercase alphabetic mode, press [↑][alpha]. As before, pressing [alpha] again takes the TI-89 out of alphabetic mode.

6   Accessing previous entries ("deep recall")

By repeatedly pressing [2nd][ENTER] (ENTRY), previously typed expressions can be retrieved for editing and re-evaluation. Pressing [2nd][ENTER] once recalls the most recent entry; pressing [2nd][ENTER] again brings up the second most recent, etc.

More conveniently, pressing ⊙ and ⊙ allows one to select previous entries and results from the history area; simply highlight the desired expression and press [ENTER]. For example, pressing ⊙⊙[ENTER] would copy the previous entry to the new-entry line, while ⊙⊙⊙[ENTER] would copy the second-previous *result*.

Once an expression is on the new-entry line at the bottom of the home screen, it can be edited in various ways. Text can be deleted (using [←] to delete the character before the cursor, or [♦][←] to delete the character after the cursor). New text can be inserted (see the previous section). One can even highlight text by pressing and holding [↑] together with an arrow key, then cut ([♦][2nd]) or copy ([♦][↑]) it to a "clipboard," so that it can be pasted ([♦][ESC]) somewhere else in the expression.

7   Menus

Certain keys and key combinations bring up a menu in a window with a variety of options. Shown is the menu produced by pressing [F2] (Algebra) from the home screen. The arrow next to "8" means that there are more options available (which can be seen by pressing ⊙ or ⊙). To select one of these options (and paste the corresponding command on the entry line), simply press the number (or letter) next to the option. Alternatively, use ⊙ and ⊙ to highlight the desired option and press [ENTER].

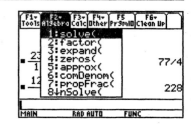

[2nd][5] brings up the MATH menu shown in the first screen on the right. For each of these menu options, the triangle ("▸") on the right side indicates that selecting that option brings up a sub-menu. Shown on the far right is the List sub-menu (option 3 of the MATH menu). Note

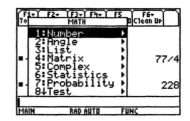

that the status line contains a reminder of how to use these menus; [ESC] can be used to exit from one level of a menu (and [2nd][ESC] would remove all menus and return to the home screen).

The various commands in these menus are too numerous to be listed here. They will be mentioned as needed in the examples.

One last comment is worthwhile, however. Some functions that may be used frequently are buried several levels deep in the menus, and may take many keystrokes to access. Worse, the location of the function might be forgotten (is it in the Algebra or MATH:Number menu?), necessitating a search through the menus. It is useful to remember three things:

- Any command can be typed one letter at a time, in either upper- or lowercase; e.g., [↑][ALPHA][=][(] [3][ALPHA][(] will type the letters "ABS(", which has the same effect as [2nd][5][1][2].

- Any command can be found in the [CATALOG] menu. Since the commands appear in alphabetical order, it may take some time to locate the desired function. Pressing any letter key (it is not necessary to press [alpha] first) brings up commands starting with that letter ; e.g., pressing [9] brings up the list on the right, while pressing [-] shows commands starting with "O."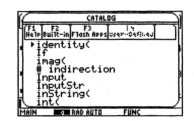

- An alternate home-screen menu bar can be found by pressing [2nd] [HOME] (CUSTOM). Pressing [2nd][HOME] again toggles back to the "standard" home-screen menu bar. It is possible to change the commands listed in this CUSTOM menu, but the process is somewhat tedious. Full details can be found in the TI-89 manual, but as an example: The command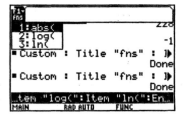

    Custom:Title "Fns":Item "abs(":Item "log(":Item "ln(":EndCustm

would result in a CUSTOM menu bar for which [F1] would produce the menu shown on the right.

8   Variables

The letters A through Z (upper- or lowercase), and also sequences of letters (like "High" or "count") can be used as variables (or "memory") to store numerical values. To store a value, type the number (or an expression) followed by [STO▸], then a letter or letters (pressing [alpha] if necessary), then [ENTER]. That variable name can then be used in the same way as a number, as demonstrated at right. These variable names are *not* case-sensitive, so "A" is the same as "a," and "CoUnT" is the same as "count."

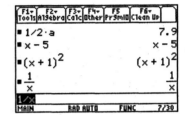

Note: The TI-89 interprets 2a as "2 times a"—the "*" symbol is not required (this is consistent with how we interpret mathematical notation). As for order of operations, this kind of multiplication is treated the same as "*" multiplication (see the screen above).

If a variable is used for which no value has been assigned, it is treated as an unknown value, and expressions involving it remain unevaluated. In the screen on the right, the variable x has no assigned value. Any variable's assigned value can be erased ("forgotten") by issuing the command DelVar ([F4][4] from the home screen menubar) followed by the variable name. All one-letter variables can be cleared by choosing [2nd][F1] (Clean Up) [1], or using the NewProb command ([2nd][F1][2]), which also clears the history area.

9   Setting the modes

By pressing [MODE], one can change many aspects of how the calculator behaves. For most of the examples in this manual, the MODE settings will be as shown on the three screens below (although in some cases the settings are not crucial). Some (not all) of these options are described below; consult the TI-89 manual for more details. Changes in the settings are made using the arrows keys and [ENTER].

 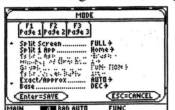 

The Graph setting can be either FUNCTION, PARAMETRIC, POLAR, SEQUENCE, 3D, or DIFF EQUATIONS. The first three of these are used for this manual. This setting determines whether formulas to be graphed are functions (*y* as a function of *x*), parametric equations (*x* and *y* as functions of *t*), polar equations (*r* as a function of *θ*), or sequences (*u* as a function of *n*). (The current value of this setting is indicated in the status line at the bottom of the calculator screen—FUNC, PAR, etc.)

The Display Digits setting can be FLOAT, FIX *n*, or FLOAT *n*, where *n* is an integer from 1 to 12. This specifies how numbers should be displayed: FLOAT means that the TI-89 should display all non-zero digits (up to a maximum of 12), while (e.g.) FLOAT 4 means that a total of 4 digits will be displayed. Meanwhile, FIX 4 means that the TI-89 will attempt to display 4 digits beyond the decimal point.

Angle can be either RADIAN or DEGREE, indicating whether angle measurements should be assumed to be in radians or degrees. (A right angle measures $\frac{\pi}{2}$ radians, which is equivalent to 90°.) Most of the examples in the text include reminders to set the calculator in the appropriate mode, in cases where this is important. The current value of this setting is indicated in the status line at the bottom of the calculator screen—RAD or DEG.)

The Exponential Format setting is either NORMAL, SCIENTIFIC, or ENGINEER-ING; this specifies how numbers should be displayed. The screen on the right shows the number "12345." displayed in each of these modes: NORMAL mode displays numbers in the range ±999,999,999,999 with no exponents, SCI-ENTIFIC mode displays all numbers in scientific notation, and ENGINEERING mode uses only exponents that are multiples of 3. Note: "E" is short for "times 10 to the power," so $1.2345\text{E}4 = 1.2345 \times 10^4 = 1.2345 \times 10000 = 12345$.

The Complex Format is either REAL, RECTANGULAR, or POLAR, and specifies the display mode for complex numbers. REAL means that the TI-89 will produce an error if an expression requires the computation of (e.g.) a square root of a negative number, and the other two settings determine whether complex results should be displayed in rectangular or polar format. These two formats are essentially the same as the two used by the textbook. **Note:** The text prefers the term "trigonometric format" rather than "polar format." More information about complex number formats can be found beginning on page 107 of this manual.

The Vector Format setting (RECTANGULAR, CYLINDRICAL, or SPHERICAL) indicates the default display format for vectors (see page 106 of this manual).

Pretty Print (ON or OFF) determines how expressions (input and output) should be displayed. The first two entries on the right were performed with Pretty Print on, and the other two were done with Pretty Print off.

The Exact/Approx setting (AUTO, EXACT, or APPROXIMATE) determines whether results should be considered to be exact or approximate. EXACT means that all decimals are converted to fractions; e.g., .9 is displayed as 9/10, and $\int(2.5)$ is $\sqrt{10}/2$. APPROXIMATE means that everything is converted to decimal form; for example, $\int(2.5)$ produces 1.5811.... With the AUTO setting, the TI-89 decides whether to display a result as exact or approximate based on whether there is a decimal point in the entry—for example, $\int(5/2)$ yields $\sqrt{10}/2$, while $\int(2.5)$ yields 1.5811.... **Note:** Regardless of this setting, pressing ◆ENTER instead of ENTER to process an entry will cause the TI-89 to show a decimal (approximate) result. (The current value of this setting is indicated in the status line at the bottom of the calculator screen—AUTO, EXACT, or APPROX.)

The other mode settings deal with issues that are beyond the scope of the textbook, and are not discussed here.

10   Setting the graph window

Pressing ◆F2 brings up the WINDOW settings. The exact contents of the WINDOW menu vary depending on the Graph mode setting; below are six examples showing this menu in each possible Graph modes. (The last two are not used in this manual, but are shown here for reference.)

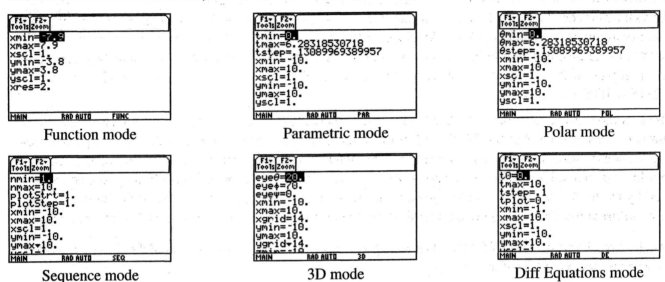

Function mode   Parametric mode   Polar mode

Sequence mode   3D mode   Diff Equations mode

All these menus include the values `xmin`, `xmax`, `xscl`, `ymin`, `ymax`, and `yscl`. When ◆F3 (GRAPH) is pressed, the TI-89 will show a portion of the Cartesian (*x*-*y*) plane determined by these values. In function mode, this menu also includes `xres`, the behavior of which is described in section 12 of this manual (page 89). The other settings in this screen allow specification of the smallest, largest, and step values of *t* (for parametric mode) or θ (for polar mode), or initial conditions for sequence mode.

With settings as in the "Function mode" screen shown above, the TI-89 would display the screen at right: *x* values from −7.9 to 7.9 (that is, from `xmin` to `xmax`), and *y* values between −3.8 to 3.8 (`ymin` to `ymax`). Since `xscl` = `yscl` = 1, the TI-89 places tick marks on both axes every 1 unit; thus the *x*-axis ticks are at −7, −6, . . . , 6, and 7, and the *y*-axis ticks fall on the integers from −3 to 3. This window is called the "decimal" window, and is most quickly set by pressing ◆F2 F2 4 (Zoom:ZoomDec).

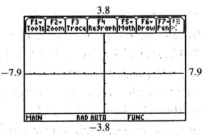

Below are three more sets of window settings, and the graph screens they produce. Note that the first graph on the left has tick marks every 10 units on both axes. The second window is called the "standard" viewing window, and is most quickly set by pressing ◆F2F26 (Zoom:ZoomStd). The setting $\mathtt{yscl} = 0$ in the final graph means that no tick marks are placed on the y-axis.

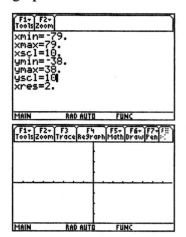
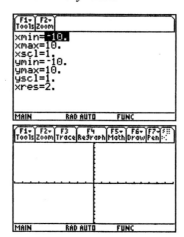

11   The graph screen

The TI-89's graph screen—that is, the portion of the screen used to display graphs, below the menu bar and above the status line—is made up of an array of rectangular dots (pixels) with 77 rows and 159 columns. All the pixels in the leftmost column have x-coordinate xmin, while those in the rightmost column have x-coordinate xmax. The x-coordinate changes steadily across the screen from left to right, which means that the coordinate for the nth column (counting the leftmost column as column 0) must be $\mathtt{xmin} + n\Delta x$, where $\Delta x = (\mathtt{xmax} - \mathtt{xmin})/158$. Similarly, the nth row of the screen (counting up from the bottom row, which is row 0) has y-coordinate $\mathtt{ymin} + n\Delta y$, where $\Delta y = (\mathtt{ymax} - \mathtt{ymin})/76$.

It is not necessary to memorize the formulas for Δx and Δy. Should they be needed, they can be determined by pressing ◆F3 (GRAPH) and then the arrow keys. When pressing ⓞ or ⓞ successively, the displayed x-coordinate changes by Δx; meanwhile, when pressing ⊖ or ⊖, the y-coordinate changes by Δy. Alternatively, the values can be found by typing "Δx" and "Δy" on the home screen; this is most easily done by pressing 2nd+15 to access the CHAR:Greek menu and type the "Δ" character, then press X or Y. This produces results like those shown on the right; the values of Δx and Δy there are those for the standard viewing window.

In the decimal window $\mathtt{xmin} = -7.9, \mathtt{xmax} = 7.9, \mathtt{ymin} = -3.8, \mathtt{ymax} = 3.8$, note that $\Delta x = 0.1$ and $\Delta y = 0.1$. Thus, the individual pixels on the screen represent x-coordinates $-7.9, -7.8, -7.7, \ldots, 7.7,$ $7.8, 7.9$ and y-coordinates $-3.8, -3.7, -3.6, \ldots, 3.6, 3.7, 3.8$. This is where the decimal window gets its name.

Windows for which $\Delta x = \Delta y$, such as the decimal window, are called square windows. Since there are just over twice as many columns as rows on the graph screen, this means that square windows should have $\mathtt{xmax} - \mathtt{xmin}$ just over twice as big as $\mathtt{ymax} - \mathtt{ymin}$. Any window can be made square be pressing ◆F2F25 (Zoom:ZoomSqr). To see the effect of a square window, observe the two pairs of graphs below. In each pair, the first graph is on the standard window, and the second is on a square window (after pressing ◆F2F25). (This changes xmin and xmax to about -20.8 and 20.8, respectively, while ymin and ymax remain unchanged at -10 and 10.) The first pair shows the line $y = x$; on the square window, this line (correctly) appears to make a $45°$ angle with the x- and y-axes. The second pair shows the lines $y = 2x - 3$

and $y = 3 - \frac{1}{2}x$; note that on the square window, these lines look perpendicular (as they should). Finally, the last pair shows a circle centered at the origin with a radius of 8. On the standard window, this looks like an oval since the screen is wider than it is tall. (The reason for the gaps in the circle will be addressed in the next section.)

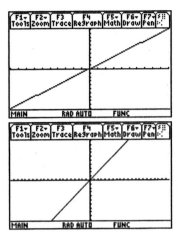 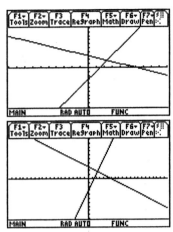 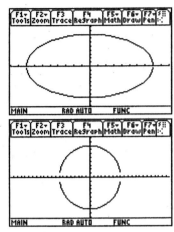

12   Graphing a function

This introductory section only addresses creating graphs in function mode. Procedures for creating parametric and polar graphs are very similar; they are covered beginning on page 110 of this manual, in material related to Chapter 8 of the text.

To see the graph of $y = 2x - 3$, begin by entering the formula into the calculator. This is done by pressing ◆F1 to access the "y equals" screen of the calculator. Enter the formula as y1 (or any other y*n*). (If y1 already has a formula, press ENTER, F3, or CLEAR first, then type the new formula.) If another y variable has a formula, position the cursor on that line and press either CLEAR (to delete the formula) or F4. The latter has the effect of toggling the check mark for that line; which tells the TI-89 whether or not to graph that formula. In the screen on the right, only y1 will be graphed.

The next step is to choose a viewing window. See the previous section for more details on this. This example uses the standard window (F2 6).

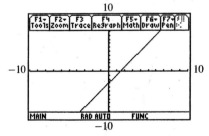

If the graph has not been displayed, press ◆F3, and the line should be drawn. In order to produce this graph, the TI-89 considers 159 values of x, ranging from xmin to xmax in steps of Δx (assuming that xres $= 1$; see below for other possibilities). For each value of x, it computes the corresponding value of y, then plots that point (x, y) and draws a line between this point and the previous one. (See also the information about graph styles later in this section.)

If xres is set to 2, the TI-89 will only compute y for every other x value; that is, it uses a step size of $2\Delta x$. Similarly, if xres is 3, the step size will be $3\Delta x$, and so on. Setting xres higher causes graphs to appear faster (since fewer points are plotted), but for some functions, the graph may look "choppy" if xres is too large, since detail is sacrificed for speed.

Note: If the line does not appear, or the TI-89 reports an error, double-check all the previous steps. Also, check the mode settings (discussed in section 9, page 85).

Once the graph is visible, the window can be changed using ⊡F2 (WIN-DOW) or F2 (Zoom). Pressing F3 (TRACE) brings up the "trace cursor," and displays the x- and y-coordinates for various points on the line as the ⊲ and ⊳ keys are pressed. (These variables—xc and yc—can also be referenced from the home screen; that is, typing xc ENTER on the home screen would show the value 2.40506....) Tracing beyond the left or right columns causes the TI-89 to adjust the values of xmin and xmax and redraw the graph.

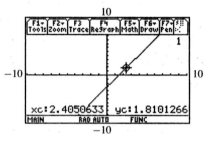

To graph the function

$$y = \frac{1}{x-3},$$

enter that formula into the "y equals" screen (note the use of parentheses on the entry line). As before, this example uses the standard viewing window.

For this function, the TI-89 produces the graph shown on the right. This illustrates one of the pitfalls of the connect-the-dots method used by the calculator: The nearly-vertical line segment drawn at $x = 3$ *should not be there*, but it is drawn because the calculator connects the points

$$x \approx 2.911, y \approx -11.286 \text{ and } x \approx 3.165, y \approx 6.077.$$

Calculator users must learn to recognize these flaws in calculator-produced graphs.

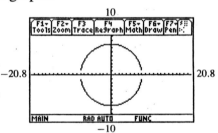

The graph of a circle centered at the origin with radius 8 (shown on a square window, with xres = 1) shows another problem that arises from connecting the dots. When $x = -8.157895$, y is undefined, so no point is plotted (that is, there is no point on this circle that has x-coordinate less than -8, or greater than 8). The next point plotted on the upper half of the circle is $x = -7.894737$ and $y = 1.2934953$; since no point had been plotted for the previous x-coordinate, this is not connected to anything, so there appears to be a gap between the circle and the x-axis. The calculator is not "smart" enough to know that the graph should extend from -8 to 8.

One additional feature of graphing with the TI-89 is that each function can have a "style" assigned to its graph. To see this style, go to the "y equals" screen and press 2nd F1 (Style); the check mark indicates which style applies to the current function. These options are shown on the right; "Line," the default, means that the calculator should draw lines between the plotted points. Complete details are in the TI-89 manual.

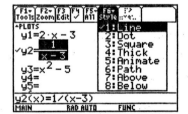

13   Adding programs to the TI-89

The TI-89's capabilities can be extended by downloading or entering programs into the calculator's memory. Instructions for writing a program are beyond the scope of this manual, but programs written by others and downloaded from the Internet (or obtained as printouts) can be transferred to the calculator in one of three ways:

1. If one TI-89 already has a program, it can be transferred to another using the calculator-to-calculator link cable. To do this, first make sure the cable is firmly inserted in both calculators. On both calculators, press [2nd][−] (VAR-LINK). On the sending calculator, use the arrow keys and [F4] to place check marks next to the programs to be transferred. On the receiving calculator, press [F3][2] (Link:Receive), then on the sending calculator, press [F3][1] (Link:Send to TI-89/TI-92 Plus).

2. If a computer with the TI-Graph Link is available, and the program file is on that computer (e.g., after having been downloaded from the Internet), the program can be transferred to the calculator using the TI Graph Link software. This transfer is done in a manner similar to the calculator-to-calculator transfer described above; specific instructions can be found in the documentation that accompanies the Graph Link software. (They are not given here because of slight differences between platforms and software versions.)

3. View a listing of the program and type it in manually. (**Note:** Even if the TI-Graph Link cable is not available, the Graph Link software can be used to view program listings on a computer.) While this is the most tedious method, studying programs written by others can be a good way to learn programming. To enter a program, start by choosing [APPS][7][3] (Program Editor:New), then specify whether this is a program  or a function, and give it a name (up to eight characters, like "quadform" or "midpoint")—note that the TI-89 is automatically put into alpha mode while typing this name. Then press [ENTER] (OK), enter the commands in the program or function, and press [2nd][ESC] (QUIT) to return to the home screen when finished.

To run the program, make sure there is nothing on the current line of the home screen, then type the name of the program or function (this name is not case sensitive). Follow this with a set of parentheses, containing any required arguments, then press [ENTER]. If the program was entered manually (option 3 above), errors may be reported; in that case, press [ENTER] (GOTO), correct the mistake and try again.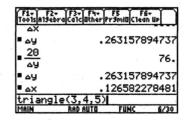

Programs can be found at many places on the Internet, including:

- http://www.awl.com/lhs—the Web site for the text;

- http://www.bluffton.edu/~nesterd—the Web site of the author of this manual;

- http://tifaq.calc.org—a "Frequently Asked Questions" page maintained by Ray Kremer; and

- http://www.ticalc.org.

Examples

Here are the details for using the TI-89 for several of the examples from the textbook. Also given are the keystrokes necessary to produce some of the commands shown in the text's examples. In some cases, some suggestions are made for using the calculator more efficiently.

We first consider examples from the text's Appendices, as the calculator techniques they illustrate are useful throughout the text.

Throughout this section, it is assumed that the textbook is available for reference. The problems from the text are not restated here, and there are frequent references to the calculator screens shown in the text.

Appendix A   Example 1   (page 434)   Solving a Linear Equation

Here is a general discussion of how to use the TI-89 to solve (or confirm solutions for) nearly any equation. We will show multiple approaches for solving the equation $\frac{1}{2}x - 6 = \frac{3}{4}x - 9$. (The answer is $x = 12$.) These procedures can be adapted for any equation, including the one from this example, or those found throughout the text.

There are two graphical methods that can be used to confirm this solution. The first is the **intersection** method. To begin, set up the TI-89 to graph the left side of the equation as y1, and the right side as y2. **Note:** Entering the fractions in parentheses—e.g., y1=(1/2)x-6—ensures no mistakes with order of operations. This is not crucial for the TI-89, but is a good practice because some other models give priority to implied multiplication. See section 8 of the introduction, page 85.

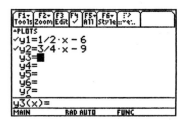

We are looking for an x value that will make the left and right sides of this equation equal to each other, which corresponds to the x-coordinate of the point of intersection of these two graphs.

Next, select a viewing window which shows the point of intersection; we use $[-15, 15] \times [-10, 10]$ for this example. The TI-89 can automatically locate this point using the GRAPH:Math:Intersection feature. Use $\odot$, $\ominus$ and ENTER to specify which two functions to use (in this case, the only two being displayed). The TI-89 then prompts for lower and upper bounds (numbers that are, respectively, less than and greater than the location of the intersection). After pressing ENTER, the TI-89 will try to find an intersection of the two graphs. The screens below illustrate these steps.

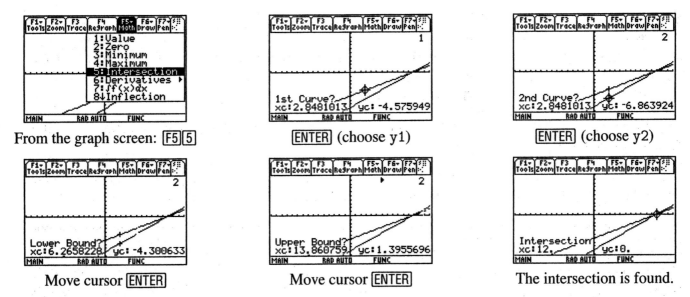

From the graph screen: F5 5   ENTER (choose y1)   ENTER (choose y2)

Move cursor ENTER   Move cursor ENTER   The intersection is found.

The x-coordinate of this point of intersection is calculated to 14 digits of accuracy, so if the solution were some less "convenient" number (say, $\sqrt{3}$ or $1/\pi$), we would have an answer that would be accurate enough for nearly any computation.

Note: An approximation for the point of intersection can be found simply by moving the TRACE cursor as near the intersection as possible. The amount of error can be minimized by "zooming in" on the graph. This is the only method available for graphing calculators such as the TI-81.

The second graphical approach is to use the x-**intercept method**, which seeks the x-coordinate of the point where a graph crosses the x-axis. Specifically, we want to know where the graph of y1−y2 crosses the x-axis, where y1 and y2 are as defined above. This is because the equation $\frac{1}{2}x - 6 = \frac{3}{4}x - 9$ can only be true when $\frac{1}{2}x - 6 - \left(\frac{3}{4}x - 9\right) = 0$.

To find this x-intercept, begin by defining y3=y1(x)−y2(x) on the Y= screen. We could do this by re-typing the formulas entered for y1 and y2, but having typed those formulas once, it is more efficient to do this as shown on the right. Note that y1 and y2 have been "de-selected" so that they will not be graphed (see section 12 of the introduction, page 89).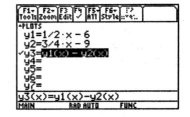

We must first select a viewing window which shows the x-intercept; we again use $[-15, 15] \times [-10, 10]$. The TI-89 can automatically locate this point with the GRAPH:Math:Zero feature. The TI-89 prompts for lower and upper bounds (numbers that are, respectively, less than and greater than the zero), then attempts to locate the zero between the given bounds. (Provided there is only one zero between the bounds, and the function is "well-behaved"—meaning it has some nice properties like continuity—the calculator will find it.) The screens below illustrate these steps.

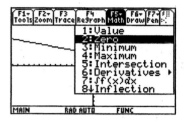

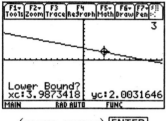

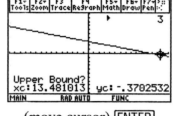

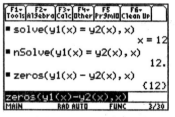

(move cursor) ENTER   (move cursor) ENTER   Here is the result.

The TI-89 also offers some non-graphical approaches to solving this equation (or confirming a solution): As illustrated on the right, the TI-89's solve and nSolve functions attempt to find solutions to an equation, while the zeros function attempts to find the zeros of an expression. As shown here, the entries make use of the fact that y1 and y2 have been defined as the left and right sides of this equation; if that had not been the case, the same results could have been attained by entering (e.g.) solve((1/2)x-6=(3/4)x-9,x). Full details on how to use these functions (all of which are found in the Algebra menu) can be found in the TI-89 manual.

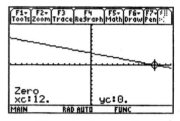

Finally, the TI-89 includes an "interactive solver," accessed with APPS 9. This prompts for the equation to be solved, then allows the user to enter a guess for the solution (or a range or numbers between which a solution should be sought). To solve the equation, place the cursor on the line beginning with x= and press F2 (Solve).

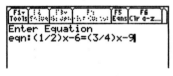

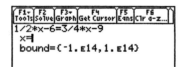

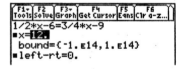

Enter equation   Specify guess, or press F2   Here is the solution.

The solver can also be used with equations containing more than one variable; simply provide values for all but one variable, then place the cursor on the line containing the variable for which a value is needed and press F2.

Note: In this example, we learned how the TI-89 can be used to support an analytic solution. But the TI-89 and any other graphing calculator also can be used for solving problems when an analytic solution is **not** possible—that is, when one cannot solve an equation "algebraically." This is often the case in many "real-life" applications, and is one of the best arguments for the use of graphing calculators.

Appendix A   Example 4   (page 436)   Using the Zero-Factor Property

---

Appendix A   Example 5   (page 437)   Using the Square-Root Property

---

Appendix A   Example 6   (page 437)   Using the Quadratic Formula

The `solve` and `zeros` functions, described above, can be used for many types of equations. The third screen shows an equation with complex solutions, for which we use `cSolve` or `cZeros` rather than `solve` and `zeros`.

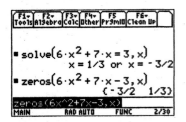 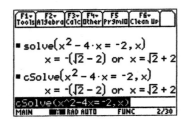 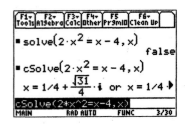

Appendix A   Example 7   (page 438)   Solving a Linear Inequality

The `solve` feature (described on page 94) will solve inequalities as well as equations. ([2nd]. types >; [♦][0] types ≤; these symbols are also found in the MATH:Test menu.) While this is much faster, students should recognize that there are some benefits to doing problems like this by hand (the "hard" way)—specifically, working through the steps shown in the text gives one insight into what is going on, while using `solve` lends little to one's understanding of inequalities.

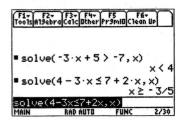

For some inequalities, `solve` fails to give useful results. However, the TI-89's `when` function can be useful in visualizing solutions (as a way to check one's algebra). In the screen on the right, note (at the bottom of the screen) how y1 was entered; this means that y1 equals 1 whenever x satisfies this inequality.

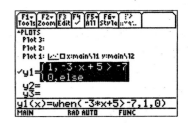

In this window, we see that y1 is equal to 1 for x to the left of 4, supporting the answer $x < 4$. Note that this picture does *not* help one determine what happens when $x = 4$; that must be checked separately.

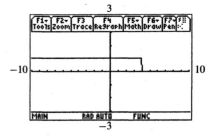

Appendix A   Example 8   (page 439)   Solving a Three-Part Inequality

The `solve` feature will only handle one inequality at a time, so we must solve the two pieces separately (as shown) and then (optionally) join them with "and" (found in the MATH:Test menu). The final output shown is equivalent to $-\frac{7}{3} < x \leq 5$.

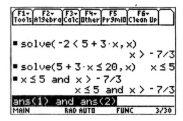

Appendix B   Example 2   (page 444)   Using the Midpoint Formula

The TI-89 can do midpoint computations nicely by putting coordinates in a list—that is, using braces ([2nd][(] and [2nd][)]) instead of parentheses. When adding two lists, the calculator simply adds corresponding elements, so the two x-coordinates are added, as are the y-coordinates. Dividing by 2 completes the task. (Note that, although the list is displayed in the history area as "{8   -4}", it was entered as "{8,-4}".)

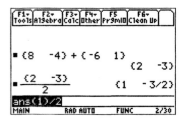

Appendix C   Example 5   (page 455)   Using Function Notation

The TI-89 is quite good with function notation. In the screen shown, y1 has been defined as the function f. It is then easy to compute $f(2)$ by entering y1(2), and (provided the variable q is undefined), we get the desired results from y1(q). The last entry shows that if q has a numerical value, the reported result is a number (rather than an expression). We can assure that all one-letter variables are undefined by pressing [2nd][F1][1] (Clean up:Clear a-z).

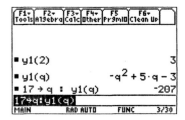

The table features of the TI-89 allow another method of computing function values. To use these features, begin by entering the formula on the Y= screen, as one would to create a graph. (The check marks determine which formulas will be displayed in the table, just as they do for graphs.)

Next, press [♦][F4] to access the Table Setup screen. The table will display y values for given values of x. The `tblStart` value sets the lowest value of x, while Δtbl determines the "step size" for successive values of x. These two values are only used if the Independent option is set to AUTO—this means, "automatically generate the values of the independent variable (x)." For this usage, Independent should be set to ASK.

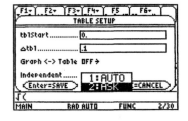

When the Table Setup options are set satisfactorily, press [ENTER] then [♦][F5] to produce the table. In the screen shown, values of y1 are shown for the three input values 0, 1, and 2. (These three input values had to be individually entered.)

To get graphical confirmation, enter the appropriate formula for y1 and graph in any window that includes $x = 2$. Press [F3] (TRACE), then press the ⊙ and ⊙ keys to change the value of x. It may not be possible to make x exactly equal to 2 in this manner (see section 12 of the introduction, page 89), but rough confirmation that $f(2) = 3$ can be found by observing that y is close to 3 when x is close to 2.

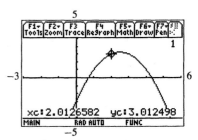

The TI-89 makes it possible to trace to any real number value for x between xmin and xmax. Simply type a number or expression (like $1/\pi$ or $\sqrt{(2)}$) while in TRACE mode. The number appears at the bottom of the window next to the xc TRACE coordinate. Pressing [ENTER] causes the TRACE cursor to jump to that x-coordinate. This same result can be achieved using the GRAPH:Math:Value command.

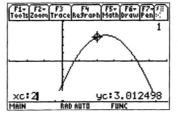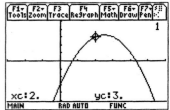

Section 1.1 Example 2 (page 4)	Calculating with Degrees, Minutes, and Seconds

| Section 1.1   Example 3   (page 4) | Converting between Decimal Degrees and Degrees, Minutes, and Seconds |

When working with angles measured in degrees, it is a good idea to select Degree mode (see section 9 of the introduction, page 85). In fact, when in Radian mode, the TI-89 automatically converts angles entered in degrees to radians, which can be confusing. (Radian measure is discussed in Chapter 2.)

In Degree mode, the conversion to decimal degrees is fairly simple: Enter the angle using [2nd][1], [2nd][=], and [2nd][1] for the degrees, minutes, and seconds symbols. The only potential snag is that, depending on the [MODE] settings, the TI-89 might attempt to display an exact (fractional) representation of the conversion, rather than a decimal (as in the first entry shown in the history area on the right). This can be overcome by either entering one of the

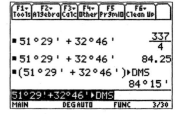

quantities (degrees, minutes, or seconds) with a decimal point, or by entering the expression with [♦][ENTER] instead of just [ENTER]. (This produced the second output in the history area.) The third entry makes use of the ▸DMS operator, which causes an angle to be displayed in degrees, minutes, and seconds, rather than as a decimal.

If the TI-89 is left in Radian mode, a little more work is needed. The MATH:Angle menu ([2nd][5][2]) is shown on the right. It includes the degrees symbol, and the ▸DMS operator (mentioned above). Not visible in this menu is option 9: ▸DD (convert to decimal degrees).

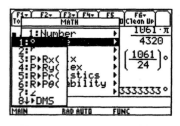

When in Radian mode, simply entering $74°8'14"$ causes the calculator to first convert interpret this as degrees (and fractions of degrees), then to convert this angle into radians. The top line in the history area shows this (not-too-helpful) result. Appending the ▸DD conversion command to this forces the TI-89 to forego the conversion to radians. The second output arose from pressing [♦][ENTER].

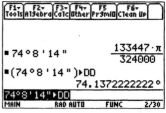

When converting from decimal degrees to DMS while the calculator is in Radian mode, be sure to enter the angle with the degree symbol ([2nd][I]); otherwise, the TI-89 assumes that the given angle is in radians, so that it converts radians to degrees before splitting the result into degrees, minutes, and seconds. Note that some rounding error was introduced in the process, causing the seconds portion of the result to be slightly different from 1.2.

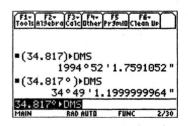

Section 1.2   Example 1   (page 10)   Finding Angle Measures

See page 92 for general information about solving equations using the TI-89. (Of course, one must use knowledge from geometry to obtain the equation in the first place.)

Section 1.3   Example 4   (page 23)   Finding Function Values of Quadrantal Angles

The alternative to putting the calculator in Degree mode is to use the degree symbol ([2nd][I]) following each angle measure; e.g., enter sin(90°) rather than just sin(90).

Unlike the TI-83 (used for the screens in the text), the TI-89 reports "undef" (undefined) when asked to compute tan 90°. Since the cotangent, secant, and cosecant functions are the reciprocals of the tangent, cosine, and sine, they can be entered as (e.g.) 1/sin(90). (This is illustrated in Figure 31 on page 27 of the text.) This reports the correct value (0) for cot 90°, even though tan 90° is undefined, and also correctly reports that sec 90° is undefined.

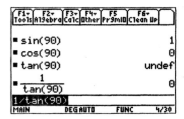

One might guess that the other three trigonometric functions are accessed with [♦] followed by [Y], [Z], or [T] (which produce, e.g., sin⁻¹). This is **not** what these functions do; in this case, the exponent −1 does not mean "reciprocal," but instead indicates that these are inverse functions (which are discussed in Sections 2.3 and 6.1 of the text). The text comments on this distinction at the bottom of page 27.

Section 2.1   Example 3   (page 48)   Solving Equations Using the Cofunction Identities

The TI-89's solve command (see page 94) gives several answers for these equations (as well as a warning that "more solutions may exist"). Aside from the answers given in the text, the other answers given by the TI-89 are correct when the trigonometric functions are expanded to work with non-acute angles.

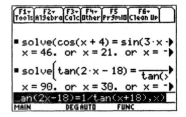

Section 2.3   Example 1   (page 62)   Finding Function Values with a Calculator

The screen on the right shows how computations like the secant in part (b) and the cotangent in (c) can be entered on a single line. These computations were done in Degree mode; in Radian mode, the first computation is done correctly, but the second requires the degrees symbol ([2nd][I]) after the angle measure.

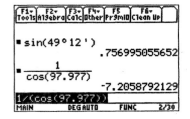

Section 2.3   Example 2   (page 62)   Using an Inverse Trigonometric Function to Find an Angle

The sin⁻¹ ("inverse sine," or "arcsine") function is ●$\boxed{Y}$, while cos⁻¹ is
●$\boxed{Z}$. The computations shown in Figure 16 in the text were done in Degree
mode; the screen on the right shows the result when done in Radian mode.
Note that in the second entry, an attempt was made to get the TI-89 to report
the result in degrees (by placing the degree symbol at the end of the entry),
but this does not have the desired result.

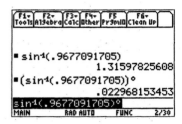

Section 2.4   Example 1   (page 69)   Solving a Right Triangle Given an Angle and a Side

Section 2.4   Example 2   (page 70)   Solving a Right Triangle Given Two Sides

For problems like these, the ans(1) variable can be used to avoid loss of
accuracy from rounding off intermediate results. Shown are computations
for Example 1; when $\boxed{\text{ENTER}}$ is pressed, the TI-89 confirms the measure of
angle B given in the text. (Be sure the TI-89 is in Degree mode. Also, recall
that the arcsine (or inverse sine) function is $\boxed{\text{2nd}}\boxed{\text{SIN}}$ —*not* $\boxed{\text{SIN}}\boxed{\text{2nd}}\boxed{\text{EE}}$.)

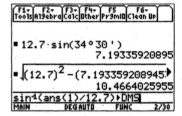

Section 2.5   Example 4   (page 79)   Solving a Problem Involving Angles of Elevation

The TI-89 can automatically locate the intersection of two graphs using the GRAPH:Math:Intersection feature.
This was previously illustrated on page 92, but we repeat the description here: With the two functions
graphed, press $\boxed{\text{F5}}$ and choose option 5 (Intersection). Use ⊙, ⊙ and $\boxed{\text{ENTER}}$ to specify which two
functions to use (in this case, the only two being displayed). The TI-89 then prompts for lower and upper
bounds (numbers that are, respectively, less than and greater than the location of the intersection). After
pressing $\boxed{\text{ENTER}}$, the TI-89 will try to find an intersection of the two graphs. The screens below illustrate
these steps; the final result is essentially the same as the screen shown in text Figure 32.

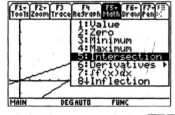
From the graph screen: $\boxed{\text{F5}}\boxed{5}$

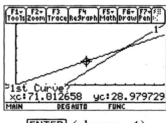
$\boxed{\text{ENTER}}$ (choose y1)

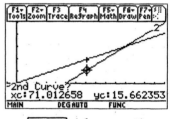
$\boxed{\text{ENTER}}$ (choose y2)

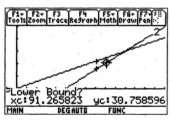
Move cursor $\boxed{\text{ENTER}}$

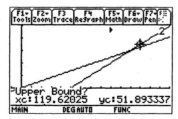
Move cursor $\boxed{\text{ENTER}}$

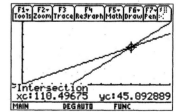
The intersection is found.

Section 3.1   Example 1   (page 95)   Converting Degrees to Radians

The number π is available as ⟨2nd⟩⟨^⟩, and the degree symbol is ⟨2nd⟩⟨I⟩. With the calculator in Radian mode (see page 85), entering $45°$ causes the TI-89 to automatically convert to radians. The screen on the right shows the two ways of performing the conversion: Letting the TI-89 do the work by using the degrees symbol, or simply multiplying by $\pi/180$.

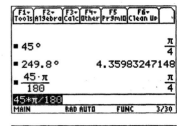

An alternative to using the degree symbol is to store $\pi/180$ in the calculator variable d (see page 85). Then typing, for example, 45d ⟨ENTER⟩ will multiply 45 by $\pi/180$. This approach will work regardless of whether the calculator is in Degree or Radian mode. (A value stored in a variable will remain there until it is replaced by a new value, or the variable is deleted or cleared.)

Section 3.1   Example 2   (page 95)   Converting Radians to Degrees

With the TI-89 in Degree mode (see page 85), the radian symbol (a super-scripted r), produced with ⟨2nd⟩⟨5⟩⟨2⟩⟨2⟩, will automatically change a radian angle measurement to degrees.

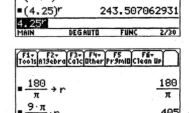

Alternatively, with the value $180/\pi$ stored in the calculator variable r (see page 85), typing $(9\pi/4)r$ ⟨ENTER⟩ will convert from radians to degrees regardless of whether the calculator is in Degree or Radian mode. (The same result can be achieved by *dividing by* the calculator variable d as defined in the previous example.)

Section 3.3   Example 4   (page 112)   Finding a Number Given Its Circular Function Value

The arccosine (or inverse cosine) function is ⟨♦⟩⟨Z⟩—*not* ⟨COS⟩⟨^⟩⟨(-)⟩⟨1⟩. Likewise, $\tan^{-1}$ is ⟨♦⟩⟨T⟩.

Section 4.1   Example 1   (page 135)   Graphing $y = a \sin x$

Note that the TI-89 must be in Radian mode in order to produce the desired graph. See page 90 for information on setting the thickness of a graph.

The graphs in the text are shown in the "trig viewing window," described on the top of page 135. The TI-89's default trig window (selected with ⟨♦⟩⟨F2⟩⟨7⟩) is slightly wider than that shown in the text; it shows x between about -3.3π to 3.3π. (If the TI-89 is in Degree mode, xmin and xmax will be ±592.5 instead of $\pm3.3\pi$.) The graph below is shown using the text's trig window, rather than the TI-89's.

It is possible to distinguish between the two graphs without having them drawn using different styles by using the TRACE feature. On the right, the trace cursor is on graph 2—that is, the graph of $y_2 = \sin x$.

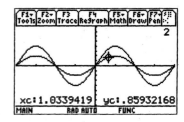

Section 4.2   Example 6   (page 151)   Modeling Temperature with a Sine Function

Given a set of data pairs (x, y), the TI-89 can produce a scatter diagram (like the points shown in Figure 19) and can find various formulas (including linear and quadratic, as well as more complex formulas like a sine function) that approximate the relationship between x and y. These formulas are called "regression formulas."

The first step is to enter the data into the TI-89. This is done by pressing [APPS][6][3] (Data/Matrix Editor), then entering a name for the "data variable" (which will contain all of the numbers for the regression). Alternatively, use an existing data variable, if there is one.

In the spreadsheet-like screen that appears, enter the month values into the first column (c1) and the temperatures into the second column (c2). If re-using an existing data variable, old data can be cleared out using Tools:Clear Editor, Utils:Delete, or Utils:Clear Column. (For the scatter diagram shown in the text, enter the temperatures twice—24 pairs of number altogether.)

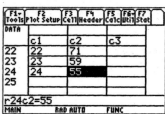

To produce the scatter diagram, press [F2][F1] (Plot Setup:Define), then choose Scatter (the default) for the Plot Type, and set x and y as c1 and c2, respectively.

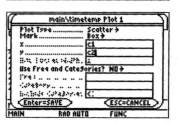

Next, check that nothing else will be plotted: Press [♦][F1] and make sure that only Plot1 has a check mark next to it. Use [F4] to turn off the check marks next to everything else.

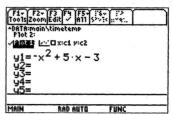

Finally, set up the viewing window as shown in Figure 19 of the text—or press [♦][F2][F2][9] (Zoom:ZoomData), which automatically adjusts the window to show all the data in the plot. This should produce a plot like that shown in the text.

Note: When finished with a statistics plot like this one, it is a good idea to turn it off so that the TI-89 will not attempt to display it the next time [♦][F3] (GRAPH) is pushed. This can be done from the Y= screen using [F4] to un-check the plot, or by pressing [F5][5] (All:Data Plots Off).

To find the regression equation, return to the data editor ([APPS][6][2])), then press [F5] (Calc). For Calculation Type, choose SinReg, and set *x* and *y* as c1 and c2, respectively.

Pressing [ENTER] displays the results of the SinReg, shown on the right—note that this computation takes several seconds. These numbers nearly agree with those shown in the text (Figure 21); the TI-89 uses a slightly different method to find this formula, so its results are not identical to those given by the TI-83.

Section 4.3   Example 1   (page 158)   Graphing $y = a \sec bx$

Section 4.3   Example 2   (page 159)   Graphing $y = a \csc (x - d)$

See section 12 of the introduction (page 89) for information about graphing functions in the Dot graph style. A reminder: sin⁻¹ ([•][Y]) is *not* the cosecant function.

Section 5.1   Example 3   (page 185)   Rewriting an Expression in Terms of Sine and Cosine

The TI-89 automatically attempts to rewrite many trigonometric expressions in terms of sine and cosine, as the screen here shows. (This will only work correctly if x has not been assigned a numerical value.)

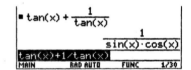

As an alternative to graphing these two functions, the TI-89's table feature (see page 96 of this manual) can be used: If the *y* values are the same for a reasonably large sample of *x* values, one can be fairly sure (though not certain) that the two expressions are equal. To make this approach more reliable, be sure to choose *x* values that are not, for example, all multiples of π.

Section 5.2	Example 1	(page 190)	Verifying an Identity (Working with One Side)

Section 5.2	Example 2	(page 190)	Verifying an Identity (Working with One Side)

Section 5.2	Example 3	(page 191)	Verifying an Identity (Working with One Side)

Section 5.2	Example 4	(page 191)	Verifying an Identity (Working with One Side)

Section 5.2	Example 5	(page 192)	Verifying an Identity (Working with Both Sides)

The TI-89 can do some simplification of this type, provided that all trig functions are entered in terms of the sine, cosine, and tangent functions. The screen on the right shows confirmation of the identities in Examples 1 and 4. For the first, the left side of the equation was entered, and the the output is equivalent to the right side. The second uses the TI-89's tCollect function, found in the MATH:Algebra:Trig menu ([2nd][5][9][9]). This can be used

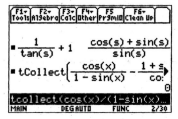

to simplify some trigonometric expressions; note that it reports that the difference between the left and right sides simplifies to 0.

The TI-89's methods of expansion do not always give the same results as those shown in these examples. Here is the output for the left side of the equation in Example 5 (which agrees with the expression found in the text). However, the TI-89 does not simplify the right side to the same expression. (With tCollect, though, it does confirm that the difference between the left and right sides is 0.)

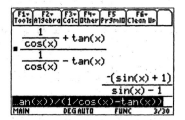

Section 5.3	Example 1	(page 198)	Finding Exact Cosine Function Values

Section 5.4	Example 1	(page 206)	Finding Exact Sine and Tangent Function Values

The TI-89 can graphically and numerically support exact value computations such as $\cos 15° = \frac{\sqrt{6}+\sqrt{2}}{4}$. Starting with a graph of y1=cos(x), the TI-89 makes it possible to trace to any real number value for x between xmin and xmax, using the GRAPH:Trace feature or the

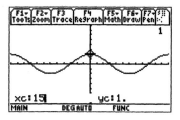
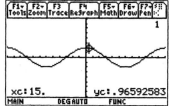

GRAPH:Math:Value command. Simply type a number or expression (like $1/\pi$ or $\int(2)$), then press [ENTER]; the cursor then jumps to that x-coordinate, and shows a decimal approximation of y.

Alternatively, a table of values like those shown here can be used to find the value of $\cos 15°$.

x	y1		
11.	.98163		
12.	.97815		
13.	.97437		
14.	.9703		
15.	.96593		

y1(x)=.96592582628907

From the home screen, we see agreement to 12 decimal places for the values of yc (which contains the *y* coordinate from the GRAPH:Trace operation), $\cos 15°$, and $\frac{\sqrt{6}+\sqrt{2}}{4}$. (Decimal points were placed in the second and third entries to get a decimal result.)

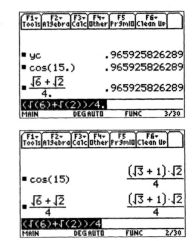

While this graphical confirmation might be more useful for other situations, it is actually unnecessary in this case, because the TI-89 knows how to use many of these identities to find exact values! Assuming that the Exact/Approx mode setting is AUTO or EXACT, the TI-89 computes the exact value of $\cos(15°)$, although it reports that exact value in a slightly different form from that given in the text.

Note that Example 1 in Section 5.6 (page 222) shows that $\cos 15°$ can also be written as $\frac{\sqrt{2+\sqrt{3}}}{2}$.

Section 5.5	Example 3	(page 215)	Verifying a Double-Angle Identity
Section 5.5	Example 4	(page 215)	Simplifying Expressions Using Double-Angle Identities
Section 5.5	Example 5	(page 216)	Deriving a Multiple-Angle Identity
Section 5.5	Example 7	(page 217)	Using a Product-to-Sum Identity
Section 5.5	Example 8	(page 218)	Using a Sum-to-Product Identity
Section 5.6	Example 1	(page 222)	Using a Half-Angle Identity to Find an Exact Value
Section 5.6	Example 2	(page 222)	Using a Half-Angle Identity to Find an Exact Value

The TI-89's tExpand function — in the MATH:Algebra:Trig menu — is more-or-less the opposite of tCollect; here we see it's expansion of the two sides of the equation in Example 3.
Note: According to the TI-89 manual, "for best results, tExpand() should be used in Radian mode." In fact, tExpand seems to do *nothing* when the TI-89 is in Degree mode.

tCollect works for some of these expressions, as well; here are it's results for Example 4.

Note, however, that there are often many equivalent ways to re-write a trigonometric expression. For Example 5, tExpand(sin(3x)) does not give the same answer as the text.

Similarly, the TI-89's version of tan 22.5° for Example 2 is not easily recognizable as equivalent to the text's $\sqrt{2} - 1$.

Section 6.1   Example 1   (page 238)   Finding Inverse Sine Values

Section 6.1   Example 2   (page 239)   Finding Inverse Cosine Values

Of course, it is not necessary to graph $y = \sin^{-1} x$ to find these values; one can simply enter, e.g., $\sin^{-1} (1/2)$. Angles are given in radians or degrees, depending on the TI-89's mode setting.

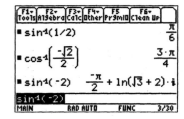

Note that if the TI-89's Complex Format is RECTANGULAR or POLAR, it does *not* give an error for the input $\sin^{-1} (-2)$, but instead gives a complex result. (A Non-real result error occurs if Complex Format is set to REAL.) This is technically a correct result, but is not appropriate for the problems in this text.

Section 6.1   Example 4   (page 243)   Finding Inverse Function Values with a Calculator

Note that the answer given for (b), 109.499054°, overrepresents the accuracy of that value. A typical rule for doing computations involving decimal values (like -0.3541) is to report only as many digits in the result as were present in the original number—in this case, four. This means the reported answer should be "about 109.5°," and in fact, any angle θ between about 109.496° and 109.501° has a cotangent which rounds to -0.3541. (See the discussion of significant digits on page 68 of the text.)

Section 6.1   Example 7   (page 245)   Writing Function Values in Terms of *u*

The TI-89 handles both of these expressions quite nicely, although it does not rationalize the denominator for (a), and only changes the second expression if the tExpand command is used.

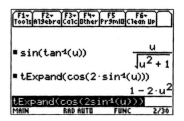

Section 6.2   Example 6   (page 253)   Describing a Musical Tone from a Graph

Section 6.3   Example 5   (page 259)   Analyzing Pressures of Upper Harmonics

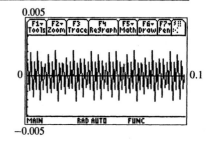

Note that the calculator screens shown in Figures 23–26 illustrate the importance of choosing a "good" viewing window. If we choose the wrong vertical scale (ymin and ymax), we might not be able to see the graph at all—it might be squashed against the *x*-axis. If we make the window too wide—that is, if xmax minus xmin is too large—we might see the "wrong" picture, like the one on the right (for Example 5 from Section 6.3): We see some suggestion of periodic behavior, but those sections of the graph that appear to be one full cycle actually consist of several cycles crammed together. The calculator cannot display sufficient detail to show the 44 full periods that fall between 0 and 0.1.

This observation—that a periodic function, viewed at fixed intervals, can appear to be a *different* periodic function—is the same effect that causes wagon wheels to appear to run backwards in old movies.

Section 7.4 Example 1 (page 307)	Finding Magnitude and Direction Angle

| Section 7.4   Example 2   (page 307) | Finding Horizontal and Vertical Components |

The TI-89 recognizes vectors entered in either of two formats:

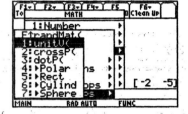

　　[*horizontal component, vertical component*] —rectangular format, or

　　[*magnitude, ∠ direction angle*] —polar format.

(The square brackets are [2nd][,] and [2nd][÷], and "∠" is [2nd][EE].) Regardless of how the vector is entered, the TI-89 displays it according to the Vector Format mode setting (see page 85); specifically, it displays the vector in component form in RECTANGULAR mode, and in magnitude/direction form for either of the other two modes. (There are, however, commands to override how the vector is displayed.)

Commands for manipulating vectors are buried several levels deep in the MATH:Matrix:Vector ops menu— [2nd][5][4][alpha][4].

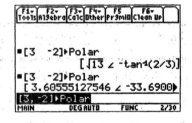

For Example 1, the vector **u** = ⟨3, −2⟩ is entered on the TI-89 as [3,-2] (but note that it is displayed in the history area without the comma). To convert to magnitude/direction form, either put the TI-89 in CYLINDRICAL or SPHERICAL mode, or use the ▶Polar command (either from the menu referred to above, or selected from the [CATALOG]), as the screen on the right illustrates. The second entry shows the decimal approximation of the magnitude and direction. (With the TI-89 in Degree mode, the angle returned is in degrees.)

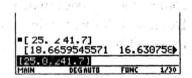

For Example 2, one can find both the horizontal and vertical components at once, as the screen on the right shows. These computations were done with the TI-89 in Degree mode, so it was not necessary to include the degree symbol on the angle. Additionally, the TI-89 was in RECTANGULAR vector display mode; if it had not been, the ▶Rect command (found in the MATH:Matrix:Vector ops menu) could be used to force conversion to rectangular format. Note that the vector is *entered* with a comma before the ∠ symbol, but *displayed* (in the history area) without the comma.

Alternatively, the TI-89 has R▶Pr, R▶Pθ, P▶Rx, and P▶Ry conversion functions, found in the MATH:Angle menu ([2nd][5][2]) as options 3 through 6. As shown in text Figures 26 and 28, these can be used to convert between the vector formats. "R" and "P" stand for "rectangular" and "polar" coordinates. Rectangular coordinates are the familiar x and y values. Polar coordinates, described in Section 8.5 of the text, are r (which corresponds to the magnitude of a vector) and θ (the direction angle).

Finally, observe that the conversion for Example 2 could be done by typing 25cos(41.7) and 25sin(41.7); which approach to use is a matter of personal preference.

Section 7.4   Example 3   (page 308)   Writing Vectors in the Form ⟨*a, b*⟩

Note how easily these conversions are done in the TI-89's polar vector format.
(The TI-89 was in Degree mode.)

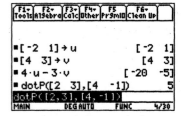

Section 7.4   Example 5   (page 309)   Performing Vector Operations

Section 7.4   Example 6   (page 310)   Finding Dot Products

Using the TI-89's vector notation, the vectors can be stored in calculator
variables (see page 85) which can then be used to do the desired operations.
Shown are the commands to compute 5(c): $4\mathbf{u} - 3\mathbf{v}$, and 6(a): $\langle 2, 3 \rangle \cdot \langle 4, -1 \rangle$.
The dotP function is in the [CATALOG] or the MATH:Matrix:Vector ops menu — [2nd]
[5][4][alpha][4][3].

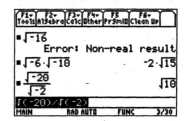

Section 8.1   Example 1   (page 333)   Writing $\sqrt{-a}$ as $i\sqrt{a}$

Section 8.1   Example 4   (page 334)   Finding Products and Quotients Involving Negative Radicands

Note that the TI-89's Complex Format mode should be RECTANGULAR (see
page 86). The screen on the right shows results from performing some
computations in REAL mode; note that this *will* work for (a), (b), and (c) in
Example 4, for which the final result is a real number, but an error occurs if
the result is complex.

The screen on the right (also produced in REAL mode) illustrates an exception
to this rule: A complex result produces no error in REAL mode if the entered
expression included the character "ι" ([2nd][CATALOG] — this is different from
the "regular" lowercase i, [alpha][9]).

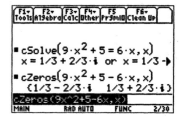

Section 8.1   Example 2   (page 333)   Solving Quadratic Equations for Complex Solutions

Section 8.1   Example 3   (page 334)   Solving a Quadratic Equation for Complex Solutions

See pages 94 and 95 for information about solve, zeros, cSolve, and
cZeros. Here is the TI-89's output for Example 3.

| Section 8.1 | Example 5 | (page 335) | Adding and Subtracting Complex Numbers |

| Section 8.1 | Example 6 | (page 336) | Multiplying Complex Numbers |

| Section 8.1 | Example 7 | (page 337) | Simplifying Powers of *i* |

| Section 8.1 | Example 8 | (page 338) | Dividing Complex Numbers |

Use [2nd][CATALOG] for the "ι" character. Note that the TI-89's Complex Format mode should be REAL or RECTANGULAR (see page 86) to produce output similar to that shown in the text. The command ▸Frac ("to fraction"), shown in the screen accompanying Example 7, is not necessary (nor is it available) on a TI-89; provided it is set to either EXACT or AUTO mode, results will be displayed as exact fractions.

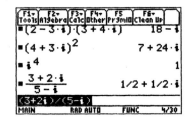

| Section 8.2 | Example 2 | (page 342) | Converting from Trigonometric Form to Rectangular Form |

With the TI-89 set to Complex format:RECTANGULAR (see page 86), or by adding the ▸Rect command to the end of the computation, the conversion to the format $a + bi$ is done automatically when a complex number is entered. Complex numbers given in trigonometric format (like those in this example) can be entered much more conveniently using either of the other two complex formats: $r \cdot e^{i\theta}$ or $(r \angle \theta)$. The entries on the right illustrate this. These computations were done in Radian mode; note the degrees symbol.

This conversion can also be done using the real and imag functions in the MATH:Complex menu, or the P▸Rx and P▸Ry conversion functions from the MATH:Angle menu. These approaches are not illustrated here, but are analogous to those shown in the text for vector conversion (see page 106).

| Section 8.2 | Example 3 | (page 343) | Converting from Rectangular Form to Trigonometric Form |

The abs and angle commands, found in the TI-89's MATH:Complex menu ([2nd] [5][5]), can be used to compute the modulus r and the argument θ (respectively). On the screen on the right, note that what is displayed as "$|-\sqrt{3} + \iota|$" was entered as "abs(-√(3)+ι)." These functions are shown on page 347 of the text.

Another approach is to use R▸Pr and R▸Pθ from the MATH:Angle menu. To do this, enter the complex number $a + bi$ as an ordered pair (a, b), as is illustrated in the screen on the right.

More convenient is the TI-89's ▶Polar command, found in the [CATALOG] or in the MATH:Matrix:Vector ops menu. This gives both the magnitude and angle at the same time, although the format in which this is displayed depends on whether the TI-89 is in Radian or Degree mode. In Radian mode (the first entry in the history area on the right), the result is displayed in the format $r \cdot e^{i\theta}$ — hence, $r = 2$ and $\theta = 5\pi/6$. When in Degree mode (the second and third entries on the right), the result is displayed as $(r \angle \theta)$ — so $r = 2$ (as before), $\theta = 150°$ for (a), and $r = 3, \theta = -90°$ for (b). Note that $-90°$ is coterminal with $270°$, given in the text.

Furthermore, with the calculator's Complex format setting as POLAR (see page 86), the conversion can be done all at once, without the ▶Polar command. (As with the ▶Polar command, the format of the output depends on whether the TI-89 is in Radian or Degree mode.)

Section 8.2   Example 4   (page 344)   Converting Between Trigonometric and Rectangular Forms
Using Calculator Approximations

Aside from entering the expression in (a) as it is shown in the text (using the cos and sin functions), one can use the TI-89's polar format $(r \angle \theta)$, but be sure to either put the TI-89 in the correct mode (Degree or Radian) or use the degrees or radians symbol from the MATH:Angle menu. Both computations on the right were done in Radian mode, and the first result is incorrect.

Section 8.3   Example 1   (page 348)   Using the Product Theorem

---

Section 8.3   Example 2   (page 349)   Using the Quotient Theorem

---

Section 8.3   Example 3   (page 349)   Using the Product and Quotient Theorems with a Calculator

---

The TI-89's polar complex format makes these computations very easy to enter.

Section 8.4   Example 1   (page 353)   Finding a Power of a Complex Number

---

While the TI-89 could be used to perform the various steps illustrated in the text (conversion to trigonometric form, etc.), note that it will compute the exact value of $(1 + i\sqrt{3})^8$ directly.

Section 8.4   Example 4   (page 356)   Solving an Equation by Finding Complex Roots

---

The TI-89's cSolve (complex solve) function can find all complex roots of a polynomial equation. The computation on the right was done in Degree and POLAR complex display mode. The visible solution $(1 \angle 72)$ corresponds to the second answer in the text; the other solutions (visible by moving the cursor into the history area and pressing ⓑ) are $(1 \angle -72)$, $(1 \angle 144)$, $(1 \angle -144)$, and 1. The two negative angles are coterminal with $216°$ and $288°$.

Section 8.5 Example 3 (page 361)	Examining Polar and Rectangular Equations of Lines and Circles

Section 8.5 Example 4 (page 363)	Graphing a Polar Equation (Cardioid)

Section 8.5 Example 5 (page 363)	Graphing a Polar Equation (Rose)

Section 8.5 Example 6 (page 364)	Graphing a Polar Equation (Lemniscate)

To produce these polar graphs, the TI-89 should be set to Degree and Polar modes (see the screen on the right). In this mode, the Y= screen is really the "r=" screen; the TI-89 allows entry of up to 99 polar equations (r as a function of θ). One could also use Radian mode, adjusting the values of θmin, θmax, and θstep accordingly (e.g., use 0, 2π, and $\pi/30$ instead of 0, 360, and 5).

For the cardioid, rose, and lemniscate, the window settings shown in the text show these graphs on "square" windows (see section 11 of the introduction, page 88), so one can see how their proportions compare to those of a circle. Note, however, that on the TI-89, these window are not square; the Zoom:ZoomSqr option can be used to adjust the window dimensions to make it square.

Press ◆F1 and set r1 to the desired expression (use ◆⌃ to type θ).

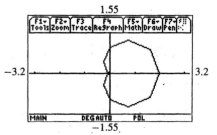

For the cardioid, the value of θstep does not need to be 5, although that choice works well for this graph. Too large a choice of θstep produces a graph with lots of sharp "corners," like the one shown on the right (drawn with θstep=30). Setting θstep too small, on the other hand, produces a smooth graph, but it is drawn very slowly. Sometimes it may be necessary to try different values of θstep to choose a good one.

The lemniscate can be drawn by setting θmin=0 and θmax=180, or θmin=−45 and θmax=45. In fact, with θ ranging from −45 to 225, the graph of r1=√(cos(2θ)) (alone) will produce the entire lemniscate. (θstep should be about 5.)

The rose can be produced by setting θmin=0 and θmax=360, or using any 360°-range of θ values (with θstep about 5).

Section 8.5 Example 7 (page 365)	Graphing a Polar Equation (Spiral of Archimedes)

To produce this graph on the viewing window shown in the text, the TI-89 must be in Radian mode. (In Degree mode, it produces the same shape, but magnified by a factor of $180/\pi$ —meaning that the viewing window needs to be larger by that same factor.)

Section 8.6   Example 1   (page 371)   Graphing a Plane Curve Defined Parametrically

Place the TI-89 in Parametric mode, as the screen on the right shows. In this
mode, ◆F1 allows entry of up to 99 pairs of parametric equations (*x* and *y*
as functions of *t*). The TI-89 will graph any pair of equations for which at
least one of *x* and *y* is selected (has a check mark next to it).

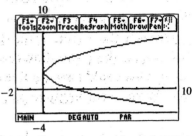

The value of tstep does not need to be 0.05, although that choice works
well for this graph. Too large a choice of tstep produces a less-smooth
graph, like the one shown on the right (drawn with tStep=1). Setting
tstep too small may produce a smooth graph, but it might be drawn very
slowly. Sometimes it may be necessary to try different values of tstep to
choose a good one.

Section 8.6   Example 3   (page 372)   Graphing a Plane Curve Defined Parametrically

This curve can be graphed in Degree mode with tmin=0 and tmax=360, or in Radian mode with tmax=2π.
In order to see the proportions of this ellipse, it might be good to graph it on a square window. This can
be done most easily with the Zoom:ZoomSqr option. On a TI-89, initially with the window settings shown in
the text, this would result in the window [−8.3, 8.3] × [−4, 4].

Section 8.6   Example 5   (page 373)   Graphing a Cycloid

The TI-89 *must* be in Radian mode in order to produce this graph.

Section 8.6   Example 6   (page 374)   Simulating Motion with Parametric Equations

Section 8.6   Example 8   (page 375)   Analyzing the Path of a Projectile

Parametric mode is particularly nice for analyzing motion, because one can picture the motion by watching
the calculator create the graph, or by using TRACE (◆F3F3) and watching the motion of the trace cursor.
(When tracing in parametric mode, the ⊙ and ⊙ keys increase and decrease the value of *t*, and the trace
cursor shows the location (*x*, *y*) at time *t*.) Figure 40 illustrates tracing on the projectile path in Example 8.
Note that the value of *t* changes by ±tstep each time ⊙ or ⊙ is pressed, so obviously the choice of tstep
affects which points can be traced.

Section 9.1	Example 4	(page 393)	Using a Property of Exponents to Solve an Equation

Section 9.1	Example 5	(page 393)	Using a Property of Exponents to Solve an Equation

Section 9.1	Example 6	(page 393)	Using a Property of Exponents to Solve an Equation

Below the calculator screen shown in the text for Example 4, the caption refers to "the *x*-intercept method of solution." This and other methods for solving equations were described beginning on page 92 of this manual. Note that the TI-89's `solve` command (see page 94) gives correct results for exponential equations like these.

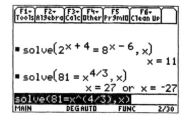

Section 9.1   Example 11   (page 397)   Using Data to Model Exponential Growth

The scatter diagram in Figure 10(a) and the "exponential regression" (ExpReg) in Figure 12 can be reproduced by adapting the procedures described on page 101 of this manual.

The two calculator screens in Figure 11 use the "intersection method" of solving equations; see page 92 for a description.

Section 9.3   Example 1   (page 414)   Finding pH

The natural logarithm function (ln) is 2nd X. For the common (base-10) logarithm, use the CATALOG or type `log` one letter at a time.

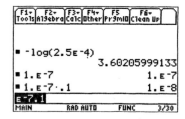

For (a), the text shows `-log(2.5*10^(-4))`. The TI-89 does not have a "10^" function, but 1 0 ^ produces the same results. This could also be entered as shown on the first line of the screen on the right, since "E" (produced with EE) and "*10^" are nearly equivalent. The two are not completely interchangeable, however; in particular, in part (b), `10^(-7.1)` **cannot** be replaced with E-7.1, because "E" is only valid when followed by an *integer*. That is, E-7 produces the same result as `10^-7`, but E-7.1 is interpreted as `(1E-7)*0.1`.

Section 9.3	Example 11	(page 420)	Solving a Logarithmic Equation

Section 9.3	Example 12	(page 420)	Solving a Base *e* Logarithmic Equation

The TI-89's `solve` command (see page 94) gives correct results for logarithmic equations like these.

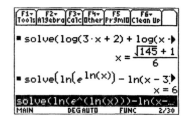